Big Data Management

The big data paradigm presents a number of challenges for university curricula on big data or data science related topics. On the one hand, new research, tools and technologies are currently being developed to harness the increasingly large quantities of data being generated within our society. On the other, big data curricula at universities are still based on the computer science knowledge systems established in the 1960s and 70s. The gap between the theories and applications is becoming larger, as a result of which current education programs cannot meet the industry's demands for big data talents.

This series aims to refresh and complement the theory and knowledge framework for data management and analytics, reflect the latest research and applications in big data, and highlight key computational tools and techniques currently in development. Its goal is to publish a broad range of textbooks, research monographs, and edited volumes that will:

- Present a systematic and comprehensive knowledge structure for big data and data science research and education
- Supply lectures on big data and data science education with timely and practical reference materials to be used in courses
- Provide introductory and advanced instructional and reference material for students and professionals in computational science and big data
- Familiarize researchers with the latest discoveries and resources they need to advance the field
- Offer assistance to interdisciplinary researchers and practitioners seeking to learn more about big data

The scope of the series includes, but is not limited to, titles in the areas of database management, data mining, data analytics, search engines, data integration, NLP, knowledge graphs, information retrieval, social networks, etc. Other relevant topics will also be considered.

John A • Satheesh Abimannan
El-Sayed M. El-Alfy • Yue-Shan Chang
Editors

Spatiotemporal Data Analytics and Modeling

Techniques and Applications

Springer

Editors
John A
Computing Science and Information Engineering
National Chung Cheng University
Chiayi County, Taiwan

Satheesh Abimannan
Amity School of Engineering and Technology
Amity University Maharashtra
Mumbai, Maharashtra, India

El-Sayed M. El-Alfy
College of Computing and Mathematics
King Fahd University of Petroleum and Mi
Dhahran, Saudi Arabia

Yue-Shan Chang
Computing Science and Information Engineering
National Taipei University
New Taipei City, Taiwan

Editor-in-Chief
Xiaofeng Meng

ISSN 2522-0179 ISSN 2522-0187 (electronic)
Big Data Management
ISBN 978-981-99-9653-7 ISBN 978-981-99-9651-3 (eBook)
https://doi.org/10.1007/978-981-99-9651-3

This Springer imprint is published by the registered company Springer Nature Singapore Pte Ltd.
The registered company address is: 152 Beach Road, #21-01/04 Gateway East, Singapore 189721, Singapore

Paper in this product is recyclable.

Preface

Spatial data, such as latitude and longitude, refer to information associated with specific locations. Having a temporal component refers to data that has a time element, like a date or a time stamp. Spatial and temporal data track changes over time and identify links across different geographical areas, respectively. Both types of data are crucial for this reason. There are a few reasons for the recent trends in spatial and temporal database management systems. One reason is the increasing amount of data that is being collected that has a spatial or temporal component to it. This data can be used to help answer questions about how things change over time or how they are related to each other geographically. Another reason is the increasing use of mobile devices and applications that need to be able to track location and time. Finally, there is a need for more efficient ways to manage and analyze this data. For instance, spatial data can be used to map the spread of disease, while temporal data can be used to track the number of cases of the disease over time. Both types of data can be combined to create a more accurate picture of the situation. A "spatial data management system" is a way to take care of data that has something to do with space. This could include data such as maps, satellite images, and GPS data. A temporal data management system (DBMS) is a system designed to manage data that has a temporal component. This could include data such as weather data, financial data, and social media data. Some common techniques used in spatial and temporal data management systems include Data compression, Data de-duplication, Data indexing, Data mining, Data warehousing, etc.

A DBMS is a system that stores data in the traditional tabular format. In contrast, a spatial and temporal data management system (STDM) stores and manages data, considering their spatial and temporal elements. Spatial and temporal data are increasingly being combined within a single database management system. This trend has been motivated by the need to enhance data management accuracy and efficiency, as well as the requirement to enhance decision-making by supplying correct and timely information. There has also been a move toward using more advanced ways to manage and analyze geographical and temporal data. This covers the application of geospatial technologies such as GIS, remote sensing, and others. In spatial and temporal database management systems, machine learning and deep learning are important because they can be used to automatically learn and extract patterns from data. This can be used to improve the performance of query processing and data analysis, as well as to make the system more user-friendly by providing predictive and recommendation capabilities.

Machine learning is used in several tasks to manage spatial and temporal databases, such as cleaning data, extracting features, and data clustering. Thus, in current scenarios, diverse intelligence models are needed to manage spatial, temporal, and moving object data. The management of spatiotemporal and moving object data models, aided by artificial intelligence, machine learning, and deep learning techniques, can automatically handle data attributes, operations, and query requirements. This book will discuss deep learning and machine learning techniques in depth. Most books on the market focus on mathematical models such as the probabilistic model, the fuzzy model, and others. Using machine learning and deep learning algorithms to analyze spatiotemporal data is necessary at present. However, this book's main objective is to encourage researchers and practitioners to share their knowledge and to incorporate recent technologies from industry and academia. The authors are expected to present their technical expertise and compare it with existing methodologies and real-time scenarios.

This book presents various spatiotemporal data analytics, modeling techniques, and their applications. It proposes recent techniques to manage spatiotemporal data using machine learning and deep learning. The applications of spatiotemporal data with recent intelligence techniques are presented in this book. The book consists of 12 chapters, each focusing on different techniques and applications. Particularly, Chap. 1 introduces spatiotemporal data, Chap. 2 presents a spatiotemporal-based recommendation system for vehicle demand, and so on. Chapters 7 and 9 present spatiotemporal applications such as e-waste and supply chain management. Chapter 8 discusses intelligent transportation system forecasting using spatiotemporal data. Chapter 10 presents renewable energy management and analysis using spatiotemporal data techniques and applications. Chapter 11 covers environmental data analytics using spatiotemporal data with different trajectory data processing and management. Finally, Chap. 12 presents various research directions and future research on spatiotemporal data.

Chiayi County, Taiwan — John A
Mumbai, Maharashtra, India — Satheesh Abimannan
Dhahran, Saudi Arabia — El-Sayed M. El-Alfy
New Taipei City, Taiwan — Yue-Shan Chang

Acknowledgments

We are deeply indebted to Almighty God for giving us this opportunity and it is only possible with the presence of God.

I express my sincere thanks to the management of National Chung Cheng University, Chiayi County, Taiwan. I wish to express my deep sense of gratitude and thanks to Springer Nature press for their valuable suggestions and encouragement.

John A. PhD

I sincerely thank my Amity University Maharashtra management for continuous support and encouragement. I thank my family members and friends for their timely support and help. I extend my thanks to Springer Nature press for excellent support and guidance.

Satheesh Abimannan, PhD

I express my sincere thanks to the management of King Fahd University of Petroleum and Minerals (KFUPM), Saudi Arabia. Also, I would like to thank the Springer Nature Press for allowing me to edit this book.

El-Sayed M. El-Alfy, PhD

I express my sincere thanks to the management of National Taipei University, Taiwan. Also, I would like to thank the Springer Nature Press for allowing me to edit this book.

Yue-Shan Chang, PhD

Contents

About the Editors

John A is currently working as a postdoctoral research fellow at AI and Sustainable Development Research Lab, National Taiwan University, Taipei, Taiwan. Pondicherry University awarded him an Undergraduate Degree in Computer Science and Engineering Discipline. He earned a postgraduate degree (M. Tech. in Computer Science and Engineering at Pondicherry University, India). In 2019, he completed his Ph.D. in Computer Science and Engineering at Manonmaniam Sundaranar University, India. His research areas of interest are real-time applications, machine learning, data analysis and prediction, and spatial and temporal data management.

Satheesh Abimannan is currently a Professor and Deputy Director in Amity School Engineering and Technology at Amity University, Mumbai. He served as a Postdoctoral research fellow at National Taipei University, Taiwan, for 1 year. He received the M.E. degree in Computer Science and Engineering from the College of Engineering, Guindy, Anna University, Chennai, and a Ph.D. degree in Computer Science and Engineering from Periyar Maniammai University. He has more than 20 years of teaching, research, and administrative experience. He received an ISTE-Young Scientist award in 2010. He has published more than 40 research articles in highly reputed international journals and visited Singapore, China, Taiwan, and Japan to present his research article at international conferences. His research interests include Deep Learning, Cloud Computing, Big-Data analytics, and information security.

El-Sayed M. El-Alfy **(Senior Member, IEEE)** is currently a Professor with the Information and Computer Science Department, Fellow of the SDAIA-KFUPM Joint Research Center for Artificial Intelligence, Affiliate of Interdisciplinary Research Center on Intelligent Secure Systems, King Fahd University of Petroleum and Minerals (KFUPM), Saudi Arabia. He has over 25 years of experience in industry and academia, involving research, teaching, supervision, curriculum design, program assessment, and quality assurance in higher education. He is an approved ABET/CSAB Program Evaluator (PEV) and a Reviewer and a Consultant for

NCAAA and several universities and research agents in various countries. He is an active researcher with interests in fields related to machine learning, computer vision, nature-inspired computing and applications to data science and cybersecurity analytics, pattern recognition, multimedia forensics, and security systems. He has published numerously in peer-reviewed international journals and conferences, edited a number of books published by reputable international publishers, attended and contributed in the organization of many world-class international conferences, and supervised master and Ph.D. students. He was also a member of ACM, the IEEE Computational Intelligence Society, the IEEE Computer Society, the IEEE Communication Society, and the IEEE Vehicular Technology Society. His work has been internationally recognized and received a number of awards. He has served as a Guest Editor for a number of special issues in international journals and been on the editorial board of a number of premium international journals, including *IEEE/CAA Journal of Automatica Sinica*, *IEEE Transactions on Neural Networks and Learning Systems*, *International Journal of Trust Management in Computing and Communications*, and *Journal of Emerging Technologies in Web Intelligence* (JETWI).

Yue-Shan Chang (Senior Member, IEEE) received Ph.D. degree from the Department of Computer and Information Science, National Chiao Tung University, in 2001. In August 1992, he joined the Department of Electronic Engineering, Ming Hsing University of Science and Technology. In August 2004, he joined the Department of Computer Science and Information Engineering, National Taipei University, Taipei, Taiwan. In August 2010, he became a Professor. He has been serving as the Chairman of the Department, since 2014, and the Dean of Student Affairs, since 2018. His research interests include information and knowledge fusion, big data analytics, cloud computing, intelligent computing, and the Internet of Things.

Part I
Spatiotemporal Data Management Techniques

Chapter 1
Introduction to Spatiotemporal Data

N. M. Balamurugan, K. Maithili, N. Revathi, R. Gayathri, and M. Adimoolam

Abstract Spatial and temporal data are generated for analytics and surveillance actions in fields like road transportation, satellite-related applications, moving action detection, etc. Various sensing technologies and social media applications have utilised spatial and temporal data. In this chapter, the study has been conducted to detail spatial and temporal data views. Later the spatial and temporal data categories were analysed alongside their merits and needfulness. Further, various spatial and temporal data models have been analysed from different aspects of the applications related to moving objects as dynamic regions and dynamic objects detection and observation. Spatial and temporal data representation techniques have been studied and compared to the highlights for further research. Furthermore, spatial and temporal data-handling methods have been studied to enrich the applications of moving objects over day-to-day societal activities alongside merits and demerits. Finally, spatial and temporal data applications have been reviewed and listed with their merits and demerits.

Keywords Spatiotemporal data · Categories of spatiotemporal data · Structure of spatiotemporal data · Representation of spatiotemporal

---

N. M. Balamurugan · N. Revathi · R. Gayathri
Department of Computer Science and Engineering, Sri Venkateswara College of Engineering, Chennai, India
e-mail: balgan@svce.ac.in; revathi@svce.ac.in; gayathri@svce.ac.in

K. Maithili
Department of Computer Science and Engineering, KG Reddy College of Engineering and Technology, Hyderabad, Telangana, India
e-mail: drmaithili@kgr.ac.in

M. Adimoolam (✉)
Department of Computer Engineering, Government Polytechnic College, Sevvapet, India
e-mail: adimoolam@gptcsevvapet.co.in

J. A et al. (eds.), *Spatiotemporal Data Analytics and Modeling*, Big Data Management, https://doi.org/10.1007/978-981-99-9651-3_1

1.1 Overview on Spatial and Temporal Data

Spatial data is the object of types like lines, points, and polygons. This data can be used for Geographical Information Systems (GIS), Environmental Studies, Global Positioning Systems (GPS), etc. [1]. Figure 1.1 depicts the objects' types of spatial data. Here all the objects are represented with a point, line or together to represent the polygon objects of different shapes. Hence, these object types can identify the objects and thereby extract information such as spatial and temporal information.

Spatial data can be represented in real time in different forms, and it is essential. Examples of spatial data are attribute data, map data, and image data [2]. Further, the spatial data can be represented as models just like data models in database management systems: vector models and raster models. Figure 1.2 shows the types of data models used for spatial data with the *x*- and *y*-axis. Here the types of data models are represented with *x*- and *y*-axis coordinates. These coordinates are used to identify the location of objects over geographically spread things. Figure 1.2 is described with coordinate information, and multiple objects can be split with these coordinates [3].

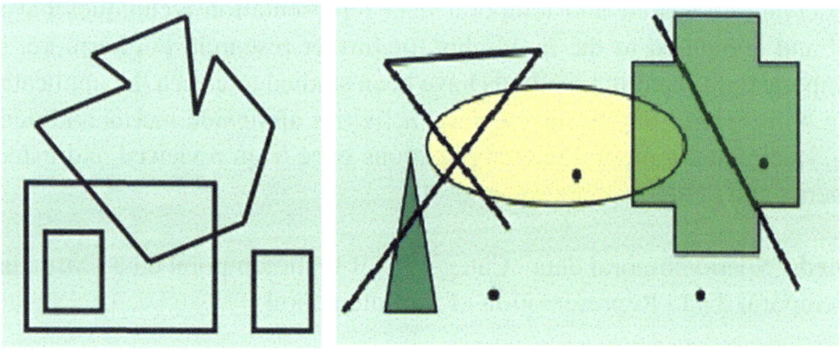

Fig. 1.1 Spatial data objects types

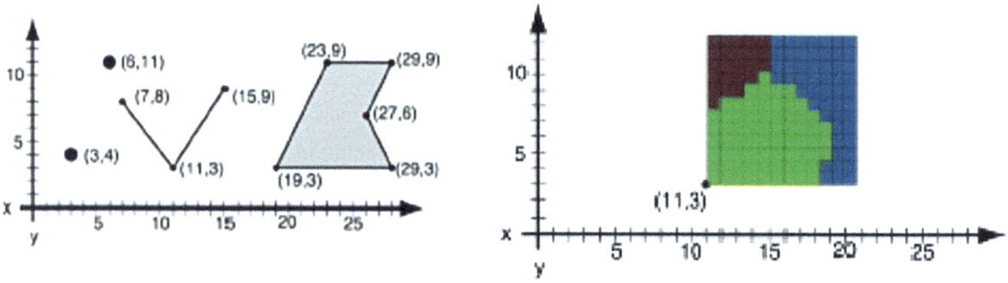

Fig. 1.2 Data model used in spatial data

1.2 Categories of Spatial and Temporal Data

The satellite-based navigation system was developed and it is called GPS [4]. It used three positions of the user namely height, longitude, and latitude alongside velocity as well as time in the format of UTC (Universal Time Coordinated). It is used to broadcast highly accurate navigation based on space-based radio navigation system and it used the pulses to users near Earth or exactly.

Any data that directly or indirectly references a definite topographical zone or scene is called spatial data. Sometimes called geospatial or terrestrial evidence, spatial data actually arithmetically characterise a corporeal entity and it is represented in a topographical synchronize organisation. It is also sometimes said as map representable components. The spatial data can be saved in several formats. Usually, these formats of spatial data use surplus than the place precise data. For some reason, these data can be squeezed individually as variable, populace, and societies [5].

Even though there are tremendous spatial data types exist in geographical regions, only two varieties are famous and those are symmetrical and terrestrial data. A data that is mapped on a two-dimensional smooth surface is known as symmetrical data. Constructed flooring plan is an example for symmetrical data. Another example will be Google Maps application. This application is essential to introduce accurate directions from symmetrical data type and this is one among simplest data for symmetrical data type. On the other hand, terrestrial data uses a sphere for information mapping. The sphere is globe Earth. Terrestrial data concentrates on the longitude and latitude relationships for identification of location kind. GPS is the best example for terrestrial data type on spatial data representation [6].

The spatial data can be in other angles based on features of three types. The first referred as point and point type can be used for location. In modern location identification-based social applications, it can be used to identify the location of restaurants, schools, etc. Second is the path spatial data type. This path spatial data type can be used to find the distance and path between one place and another [7]. The third data type is an area of spatial location. This type can be used to find the region area and use a polygon. These data types' speciality is that one data type can be transferred into another. For example, the point data type can be converted to the area by aggregating the values of all points. Ultimately, these data types can be analysed for different tasks and techniques among all data types. Figure 1.3 depicts examples of feature-based spatial data types. Around three types of this feature-based spatial data category exist [8]. The first type is a point, and locating the places over the universe is essential. The second type is the path. This path data type is used to find the path between two places. The third data type is area. This area data type is used to find the boundary of the regions [9].

The Oracle database software has introduced geometries of spatial data types to use the spatial data in real time. It has eight data types and represents SDO_GTYPE. All eight spatial data types are listed in Table 1.1 [10].

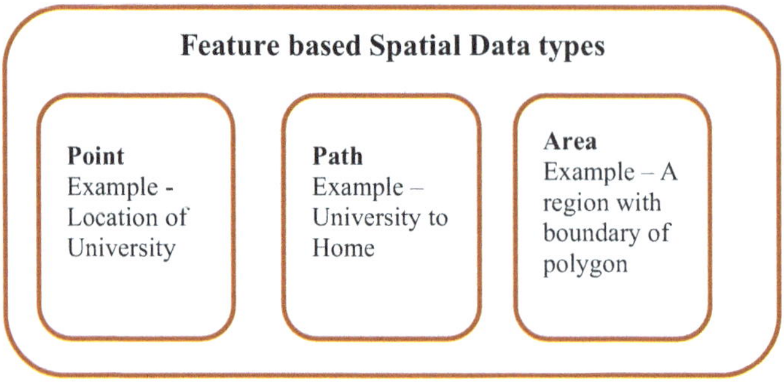

Fig. 1.3 Feature-based spatial data types

Table 1.1 Spatial data types present in Oracle database

Geometry	Description	Value
Unknown_Geometry	Ignorance of spatial data	00
Point	Used for single point element	01
Line or curve	To represent the single linear line or curve or both	02
Polygon	To present a polygon element with or without another polygon	03
Collection	To present the multiple elements	04
Multipoint	It represents one or more points	05
Multiline or multicurve	It represents single or more lines of string	06
Multipolygon	To represent the multiple polygon	07

A peculiar data is composed of spatial notions and temporal perceptions together and it is called geospatio-temporal data. Usually, spatial and temporal aspects are used to represent real-world objects. It has been represented as composite and multidimensional. On the other hand, traditional databases may not be efficiently succeeded and sustenance. This is the reason for storing the focussed databases with Relational Data Base Management Systems (RDBMS). While storing and meting out composite real-world entity, the spatial and temporal dimensions are used. The data type which is used to store and query is called a spatial database. The user queries about the (geo)spatial entity are represented here as reality. Example for geospatial data type is moving cars, roads, and cities (sometime stationary objects) [11].

1.2.1 Georeferencing and Geocoding

Georeferencing: It is a technique which is used to bind a raster picture or a vector coverage. Usually this is a coordinate stratagem and synchronizes. Sometime it is also carried out for warping/rubber-sheeting, translating, and transforming. Specifically, the functional values are relative to a variety of spatial data. These spatial data are locations examination and intersections of road. Georeferencing can

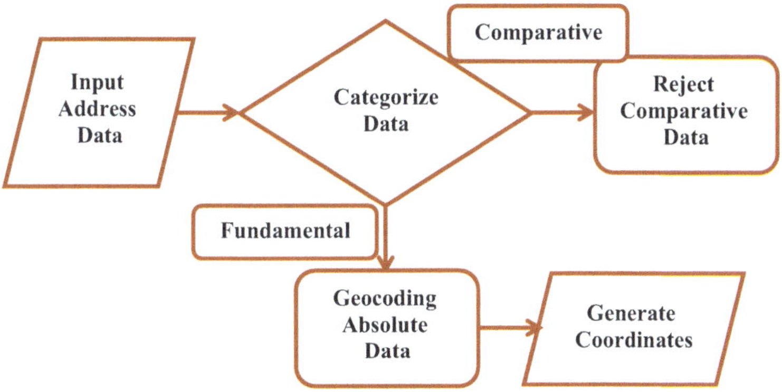

Fig. 1.4 Geocoding processing steps for find address data

now and then additionally be known as rectification or georectification interchangeably, while in some contexts, georeferencing is regarded as assigning a spatial reference solely and coordinates to the image. Rectification is the transformation and resampling of the photograph to cast off distortion (as in orthorectification). In ArcGIS, georeferencing is transitory (on-the-fly transformation of the supplied image), while rectification is everlasting (creating a new resampled photograph given a georeferenced raster layer) [12].

Georeferencing entails becoming a picture of the Earth based totally on matching up visible elements of the photo with their recognised location. Geocoding entails changing some human place name or label into coordinates. Often this is finished heaps of gadgets at a time. Input: Imagery, image manipulate points, matching geospatial manipulate points. Output: Image with facts connected displaying how it matches onto the globe. Figure 1.4 illustrates the geocoding process [13].

Georeferencing and geocoding are important aspects involved in geospatial analysis. The fitting data for the real world uses georeferencing and geocoding for detecting apposite synchronizes. Georeferencing essences towards handover data synchronizes for either vectors or raster. For accurate model this georeferencing helped especially for forming the planet's shallow. On other hand, geocoding is used to afford location descriptors and address [14]. Altogether both georeferencing and geocoding are able to collect any location information including countries, states, and cities.

1.3 Data Model Representation from Spatial and Temporal Information

The data representing the reference of location is said to be spatial data. The genuine objects presented over exact location in the factual world with the guidance of spatial data. Both the spatial and non-spatial represented alongside the aspects of location information as spatial data almost everywhere to identify the location.

These location objects include a river location with their name, start and end location information [1, 15]. To enhance spatial data, the spatial data model is further developed and are categorised into vector data (entity or object) model and raster data (field) model.

1.3.1 Vector (Entity) Data Model

It is a model used to represent the entities like storage of geospatial data (location information). Graphical information of every entity of the real world are represented for vector data model on spatial data. Majorly vector data model used points, lines, and polygon entities. The first entity is used to represent points and the needs of points are used to create lines. Further the points are enhanced as lines to find the distance between two locations. Later these lines are connected to find the areas with polygon categories. Vector data model regularly characterises the generality of entity or matters. This representation especially used over the surfaces of planet. This data model is also able to store the shapes as shape files with extension (dot shp) files [16].

1.3.2 Raster (Field) Data Model

Field is the information in format of a pixel for representing as raster value. This pixel is actually represented as grid format for a field. This Raster is supposed to be measured with a unit and it is anything like specific entity element's colour value or information. Usually, any entity imaginary is the raster value. With respect to spatial data for analysis, raster values of image can be represented as ortho-image of any place image. Usually the aerial or satellite device-based images are used for representing the raster data model (field data model) [17].

Both raster data model and vector data model are represented with attributes of image. Usually, the non-spatial data or additional information of images are related with their respective attributes. The spatial data is the attributes with any count based on resolution about location information. Some examples are photographs of entity, historical information, and a map of anything and it is necessary information for the indeed in the deed [18].

Especially, spatial data are represented uniquely and it has property that it is tangible and also it has characteristics for location and their concreteness. Usually, this raster and vector data models are coined with entity or object models. The geometry or geographic entities are framed with a shape as coordinate systems. As discussed earlier, the data types such as polygons, points, and lines are importantly used entities of geometry or geographic shapes. Every shape of entity is supposed to be represented with latitude with y axis, longitude as x axis, minimum distance of line and more than three points, respectively, for points, lines, and polygons. The beauty of this data types is interrelated since multiple points represented a line and multiple lines are represented as polygon [19].

The concept is that unidentified shapes are represented with field model and it has location information too. The specific values of field model are represented for spatial data here. It has grid format for spatial image and it has mapping function for grid representation including specific point. For example, the spatial representation is the mapping function of specific value of a globe or earth representation [2, 20].

1.3.3 Data Structures for Spatial Data

The spatial data for a database storage, a traditional 2-D indexing strategies are applied and it is unsuitable for accessing and retrieving from a database for spatial data collection. The reason is that spatial data accessing speed are increased and later the entity is retrieved for the processing and storing in a database. The information such as positions are stored over a spatial database and it has some futuristic essentialness. For example, geospatial information for a location storage are used some fields for data structures of spatial data. It is important that for storing minimum information, the two-dimensional information supposed to be stored even if it as point data type. Nowadays, even robotics and computer graphics are using the spatial data structures to store their respective works. Example processing data structure collectively uses the spatial database are quadtrees, K-dimensional (KD) trees, and R-trees as spatial data structure representation [21].

1.3.3.1 Quadtrees Data Structure

One of the hierarchical data structures used in spatial database is a quadtree. Since tree data structure used to store non-linear fashion, this tree has set the level correspondence and the advantage of this correspondence is that it can help for refinement when even the situation rose in presentation of spatial database. The mechanism of quadtree work nature is that it has internal nodes and every internal node introduced four children nodes exactly. This quadtree uses recursive function for splitting or subdividing four quadrants or regions almost all the time. Applications to application the leaf node data are associated. It has feature that information of objects has to be stored as spatial information unit of interest [22].

Based on data type such as point, polygon, and lines, the quadtrees may be confidential, including areas, points, lines, and curves. It uses tree independency for representing the order of data and further processed including curves. They are classified as one of data type as point-region quadtree (PR quadtree) internally in quadtree subordinate data type. Usually, the PR uses two-dimensional data illustration including collection of points. It has region and each region was deposed as two-dimensional points including four equal quadrants and their sub-quadrants. This process will be continued every node supposed to be having multiple point's

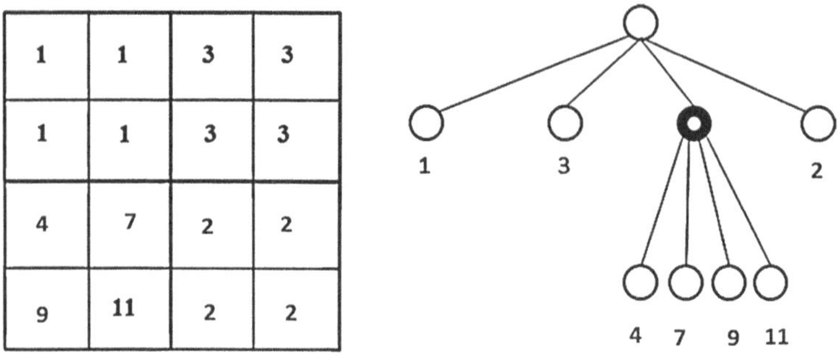

Fig. 1.5 Illustration of quadtree data structure example

under leaf node. Henceforth, the PR quadtree is the study of four-way branching it is further called 4-ary over tree data structure and their tree shape. Figure 1.5 depicts the quadtree example here. The general rule is that if the quadrant cells have the same value, they are stored as a single node leaf node without subdivision. For example, quadrants 1, 2, and 4 have the same values 1, 3, and 2, respectively. Here, all those quadrants are represented as a single-leaf node. But in quadrant 3, all cell values are different. Here quadrant 3 has 4, 7, 9 and 11. Hence these values have subdivision, as in the right side of Fig. 1.5 [23].

1.3.3.2 KD Trees

It is data structure used with KD space organized format. The KD has K value and it may be expanded to huge number. Naturally, binary tree representation is the actual format of KD tree. Every point of KD tree uses the K points. The beauty of this KD represented tree is that the non-leaf node in the tree is supposed to be act as hyperplane. Since every hyperplane is divided into space of two halves. The property of hyperplane is that it is perpendicular to axis whatever chosen. The search algorithm used to find the point is the K nearest neighbour algorithm of machine learning concept. Later the very nearest points of tree are chosen as closest to KD tree finding mechanism [24].

1.3.3.3 R-Trees Data Structure of Spatial Data

Based on B-tree representation, R-tree was introduced to present the index-based spatial data. Later the advancement of this R-tree effectively used to handle spatial data even better than KD tree spatial data representation technique. It introduced spatial access methods including multidimensional indexing of spatial data. The geographical as well as any polygons was famously used for access method of spatial data. The advantage is that it controlled overlapping very minimal and represented with minimum bounding rectangle (MBR). The reason behind this MBR is that R-tree always used to balance the tree height. Hence the input cost will be very minimal to store spatial data [25].

Properties of R-Trees

- Single root used alongside internal and leaf nodes.
- The largest region is act as the root node and pointer used to represent for any special domain.
- The child nodes are controlled by the parent nodes, children node always overlap with parent nodes.
- The MBR information will be handled by leaf node and it has representation to the current object.
- The region under consideration for MBR.

R-trees are faster than quadtrees for nearest neighbour queries while for window queries. On the other hand, spatial index creation in quadtrees is faster as compared to R-trees [26].

1.3.4 The GIS Software

Geographic information system (GIS) software transforms spatial data into information. ArcGIS, QGIS, TatukGIS, FalconView, GeoDa, Golden Software Surfer, ILWIS, SAGA GIS, GeoDa, and Bentley Map software are widely used for spatial data analysis [27].

1.4 The Spatial Data Types and Their Applications

Usually the point-referenced, areal, point patterns and network data are represented by spatial data access by data scientist. These types application oriented spatial data used extensively. The point-referenced spatial data is used to measure continuously across space. The famous spatial point instances are GPS, fixed devices, and high-resolution satellite. The major applications of point instance data are used to find the unsampled locations and helpful frequently. The predetermined group of locations was used the data type is called the areal type of spatial data and it contains information about delineated boundaries. Next the boundaries utilized for arbitrary. The utilized example is that it can be used to represent the administrative units. The units of these types include districts, regions, counties, etc. otherwise it defined the regular grid. Later this type data can be altered according to the raster data. The example applications are locations of area, spatial smoothing, model inference, and prediction over the unsampled data category [28].

Further the application extended towards the point patterns. The information for this point pattern include incident of location and purely every incident is random in nature. The major role of point incidents are used clustering or inhibition techniques for the observations and this situation based on point information of spatial location. Finally, the network type of spatial data type used the order set of points about group of straight-line information. Once again point's information details the group information. The examples of networked data type in spatial data are mobile and electronics device based-Internet including intranet. The major application of this networked data type is that it can be used in route optimization and spatial data analysis of networked devices [29].

1.5 The Applications of the Spatial and Temporal Data and its Models

1.5.1 Spatial Planning

There is a process to identify the patterns and spatial planning. For example, identification of urban growth and expansion's pattern are essential for spatial planning. If the planning of spatial information is correct then it is good to develop the areas very well as per pattern and spatial planning. The elements to be considered and it is essential for successful construction [30]

1.5.2 Agriculture Special Information

For the agriculture identification soil data are very needy to determine the crop types to be cultivated and harvested, since the crops are cultivated based on the locality and their specification. Other factors are nature of used soils. Later, this information is very needy for quick processing of agriculture for the farmers. Thus, the farmer applies their own harvesting practices. The processes like soaring and environmental standards can be improved based on agriculture spatial information alongside geospatial data.

1.5.3 Navigation Information

In today's world, no quick movement of people happens without the web navigation maps. This web navigation information is healthy to improve the movement of the public. Navigation of web maps has to be updated quickly as well as appropriately and these quick improvements are used geospatial data.

1.5.4 Telecommunication Information

Another advantage of geographical and spatial data are used in telecommunications. This telecommunication uses the construction improvement in networks. Telecommunication can be sued to track people and objects both licensed as well as unlicensed manner. Also, all web- based online shopping and enterprises system uses telecommunication information. Recently, 5G mobile network was established and it has used spatial as well as geospatial data for their accurate network construction world wid.

1.5.5 Transportation Information

Nowadays, GIS used everywhere in transportation for shortest route to avoid traffic. This transportation-based GIS helps to identify the alternate option of the current traffic, dangerous road and optimal location information. Using GIS, it is easy to identify source and destination both for road or rail track transportation.

1.5.6 Planning and Community Development Information

The subcategories of spatial data type are called geospatial data was utilized for solving the planning and developing the community of any location around the world. The usages are tremendous. The sophistication of such a applications are more with the guidance of geospatial data. The advancement and intelligences are the challenges of GIS as per the strategy plan.

1.5.7 Natural Resource Management Information

Nowadays the recent advancement of geospatial data applied in forest monitoring system for identifying the rare animal movement as well as automatice forest fire accidents. On other hand, water distribution is majorly uses the geospatial data to balance the usage and needs. Though which the environment is monitored as per distribution of natural resources.

1.5.8 Estimate Flood Damage Information

For public sector and good governance of government, natural calamities like flood damage information are essential and the resource damages by natural disaster was also measured with geospatial data. The impact of natural disaster and relief information also directly collected with geospatial data.

1.6 Spatial Data Mining

A longitudinal databank saves a huge quantity data of longitude and latitude. Usually, these data comprise remote sensing as map-based data of GIS records. Sometimes, VLSI chip format data is also used for geospatial information. The various points of spatial database used and it can be differentiated from relational databases. Spatial data elevate distance information of object and topological information. Typically, equipped objects used erudite and multidimensional longitudinal indexing. It has also used constructive access as well as spatial statistics. The techniques used for spatial data and it regularly require longitudinal intellectual alongside geometric totalling, and longitudinal understanding drawing procedures. Longitudinal evidence taking out to extracting knowledge for represented spatial relationships. This information including different fascinating patterns and it has no longer explicitly saved in longitudinal databases. Next the unification method applied for statistics mining alongside longitudinal catalogue machineries. Here the spatial records were used to discover longitudinal associations and interactions amongst longitudinal and non- longitudinal forms. The establishment of spatial understanding databases used for reorganises the longitudinal databases and sometime optimise spatial queries [31].

It is predicted to have considerable functions in geographic statistics systems including marketing, far-flung sensing, and photograph catalogue investigation. Later the same system used scientific imaging, direction finding, visitor governor, and conservational trainings. These techniques further used many areas and it was used spatial records. A central assignment to latitudinal records pulling out is the investigation of environment-friendly longitudinal information drawing out strategies due to the massive quantity of longitudinal statistics and the problem of latitudinal records sorts and spatial get right of entry to methods. Statistical altitudinal records evaluation has been a famous strategy for inspecting altitudinal statistics and exploring geographic information [32].

The period geostatistics is regularly related to non-stop geographic space, whereas the period spatial data is frequently related to discrete space. One usually considers statistical independence among extraordinary data areas in a statistical

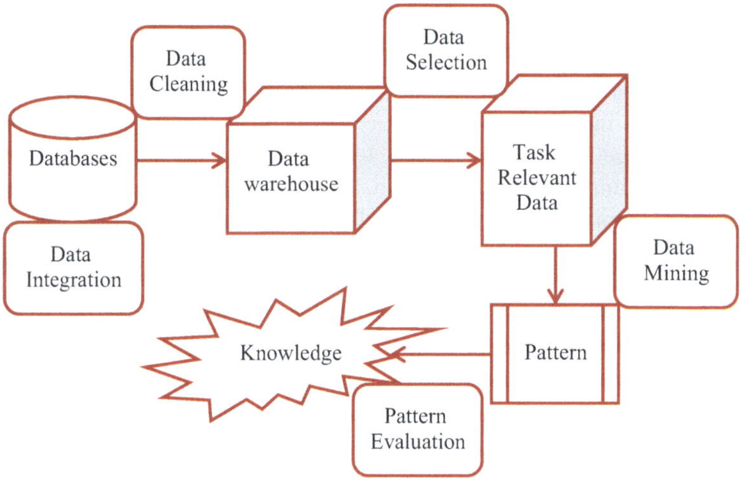

Fig. 1.6 Spatial data mining process steps

mannequin that manages non-spatial records. There is no such separation amongst spatially allotted documents because, truly, spatial objects are interrelated or extra precisely spatially co-located, in the experience that the nearer the two objects are placed, the greater probable they ship the identical properties. For example, herbal resources, climate, temperature, and monetary conditions are probable to be comparable in geographically intently placed regions [33].

A property of shut interdependency throughout the closeby area leads to the thinking of spatial autocorrelation. Based on this notion, spatial statistical modelling strategies have been successfully developed. Spatial information mining will create spatial statistical evaluation strategies and lengthen them for giant quantities of spatial data, emphasising effectiveness, scalability, cooperation with database and statistics warehouse systems, better consumer interaction, and the unearthing of new categories of acquaintance. Hence spatial data mining is initially a data cleaning process at the data warehouse, and task-relevant data is taken. This task-relevant data is further data-mined to identify the pattern. The pattern is evaluated to represent the knowledge illustrated in Fig. 1.6 [34].

1.6.1 Spatial Data Management Information

It is a technique used for retrieving information about spatial data and organizing with a Graphical Data Space (GDS) for positional information including natural locations.

1.6.2 Geospatial Data and Internet of Things (IoT)

Nowadays, IoT based GIS play inevitable role in human understandable and massive information including maps for geospatial information. For visualize the instance data of geospatial information, IoT has to be connected and get the live data for further human movement in map. Sophisticated spatial queries are applied for getting instant mapping information before human move from source to destination. The intelligence algorithms where inbuilt in a IoT devices for map navigation and data are gathered and unconnected whenever not required instantly. In general, IoT acts as data communicator in reality for geographical information and sometimes it is said to be data market. Whenever the data used as real time, the delivery value of IoT must be incorporated with geographical information. These data are two categories such as consistent and real-time longitude data. Further this data can be analysed with analytical tool including statistical information. The visual pattern gathered from IoT-based map setup uses the pattern and human can identify it. These effective visual patterns help for the intelligence to provide control mechanism including decision-making process on map information. These kinds of activities are nowadays applied in Google Mapping procedures. The collected data of geographical and geospatial are stored in common storage and cloud and cloud computing technology plays inevitable role. Here in cloud also geospatial data are measured and analysed in relevant application. The redundancy is also calculated and mapped for same spatial information geographically distance manner after sorting the comprehensive data storage. This ability is helping the cloud storage authority to move further with anywhere and anytime passion. The large-scale data of geospatial type are handled further and it has added up scale mechanism. The high-volume geospatial data further investigated with historical facts. The fruitiness this mechanism enhance the insight out of geospatial data including real time geographical data representation. The mechanism of IoT based geo spatial data are having some Key competencies and they are included as follows [4, 32]:

- Connect: It is the protocol used to interact with real-time scenario for streaming geospatial data with IoT equipped fashion and further this data can be analysed statistically from anywhere anytime passion. The reason for connect component is that I can avoid the visualize geospatial data without commuter delay.
- Analyse method: The actual gather visualized geospatial data are evaluated for quicker processing when the geospatial data are stored in cloud.
- Alert and actuate control: The pessimistic and feasible geospatial data sometimes not relevant to analyse the method. In such cases, the evaluation has to alarm the unfitted geospatial data and report can be generated for all stakeholders involved in that control mechanism.

1.6.3 Temporal Data

The representation of a state in time is called temporal data. Many applications generate temporal data which is analysed to predict weather, traffic, demographic trends and to monitor environmental variables, etc. Also, in medical field the temporal data of patients are used in disease prediction, expense prediction, and disease treatment. The temporal data is very important as it is used to understand the changes in the pattern of development or behaviour over time. The data which are used series of primary data types including generally numerical values for temporal information gathering. Since temporal data are used to provide the knowledge about temporal nature and beneficial information of time [33].

1.6.4 Applications with Temporal Aspects

The fields like accounting, finance, data warehousing and academic information use temporal database. Also, temporal data are used in GIS-based system for some control-based application. Nowadays, some more applications like law, medical and reservation systems including terrestrial, water, and air use temporal data [34].

1.6.5 Temporal Data Mining on Gathered Information

There are processes used to gather temporal data for data mining. The gathering processes include implicitly, non-trivial, and potential essentialness of temporal data. The essentialness of temporal data and their respective mining are used for pattern analysis including avoidance of unexpected data information. Sometimes, sequential data are measured for temporal information with data mining, which may be false sequence with higher-order hidden relationships. These data mining about temporal database are further investigated for time component in every geospatial information entity. The motto was altered towards temporal information including mapping information. Minding service are important characteristic of temporal data for real value information of time independent data. This data can be further investigated with some machine learning (ML) algorithms including supervised, unsupervised, and reinforcement learning for temporal behavior analysis. Recently, many researches utilized the ML for temporal investigation. The sixth sense of database are used temporal database, which are used time related information.

The major process in temporal data mining is to be processed time information irrespective of spatial information as a series. The multiple time points are measured for same set of attributes irrespective of spatial information for any visual dataset. The time attribute here plays an inevitable role for temporal data mining process. Further pattern matching method was called to detect the time-dependent geospatial information. Nowadays many temporal data mining projects introduced including spatial data as well as geospatial data for their required data and respective applications. The time values are inserted in space information collectively called temporal space measurement. The knowledge is discovered for time series calculation on visualized pattern of temporal data. The pattern of interest is also used to measure this king of temporal data mining in specific application under specific manner. Finally, temporal data mining is the complete process to identify the temporal data pattern alongside time information of visual pattern. Here temporal information are measured. The applications such as temporal data mining are used in locating the vehicle information as well as distance travelled between one time point to another time point including real object movement identification systems. Final composite of this temporal data mining is that it can measure the time limit process of spatial temporal data for comparison as well as temporal data categorization. Further mining can be extended for classification time dependent object, pattern analysis, clustering analyzing including association rule [3, 23].

1.7 Categorization of Temporal Data in Temporal Database

The categorization of temporal data is further classified as tri-temporal, bi-temporal, and uni-temporal [13, 32].

1.7.1 Tri-Temporal

It is used for supporting three axes based on time. The axes are valid time, transaction time, and decision time.

1.7.2 Bi-Temporal

It is the temporal information collected to support two axes. The axes information depend on time and they are valid time and transaction time (or) decision time.

1.7.3 Uni-Temporal

It is the temporal logic in which it uses one axis and times are dependent sometimes. The time information here used either validating the range or the system time range, that is, a valid time.

1.8 Representation of Temporal Data

The granular time series, deep learning-based temporal data, and Self-Supervised Representation Learning (SSRL) temporal data representation methods and these are discussed as follows [21, 28]

1.8.1 Granular Time Series

In granular time series modelling the time series data is represented as information granules with different levels of abstraction. The information granules are formed by technique like fuzzy clustering. The granular time series computing does the analysis of time series data grounded on logical meting out of evidence particles and appropriate instruments for understanding of logical oriented information granules. The main challenge here in granular time series analysis is the good representation scheme, accuracy, and interpretability of the time series information in the data collected. The information granules are formed by two ways:

1. Using a predefined assortment of evidence particles by experts.
2. On experimental data basis. In both methods, the membership for the fuzzy sets is fixed by using 'C' number of material morsels for both plenty and variation of fullness. The widely used experimental data information granules formation is fuzzy C-means clustering technique in which the membership to clusters are done analytically.

Once information granules are formed it is handled by appropriate models like relational modelling to capture the changing aspects of stretch succession data which is interpreted correctly by preserving the well-defined semantics of membership that is logically expressed. Once the prediction models work on information granules is done the next step is degranulation where the prediction results in the space of information granules are mapped back to the space of real numbers.

1.8.2 Deep Learning for Temporal Data

In applications like maintenance of electronic health records the handling of temporal data is really tough. It is because two kinds of data are involved in electronic health records. The structures data includes diagnoses, prescriptions, vital signs, clinical tests and procedures, whereas the unstructured data includes clinical notes, medical images, physiological signs, etc. The multiple sets of repeated variables of medical data are not efficiently handled by conventional machine learning regression models. Even the traditional statistical methods only capture the single value aggregated statistics like mean, median, mode from the input time series data which results in loss of valuable sequential information. To capture the dynamics of temporal data the deep learning-based architectures are used such as long short-term memory (LSTM), generative adversarial networks (GAN), gated recurrent unit (GRU), recurrent neural networks (RNNs), and convolution neural networks (CNNs). The main difficulty in handling temporal data in electronic health records includes varying time intervals, sparse data, and heterogeneous data [1, 4, 8, 31].

1.8.3 The SSRL Representation

To handle the time series data generated by different sensors in various sensor-based applications the self-supervised representation learning is used. This SSRL time series data representation eliminates the need for data annotation that is widely used in training of the data. This greatly reduces the difficulty and cost involved in acquiring data with annotation. The SSRL time series data representation uses discriminative pre-training of representations based on the superintendent indicators that is obtained from the underdone records. It also enables multimodal self-supervised learning for temporal data in the field of natural language processing and computer vision. To eliminate the domain experts in data annotation to make in self-learning process the SSRL approach works good on unlabelled data which works on the sequence of 'Observe, Act, Observe and Inference' procedure. The important factor in SSRL approach is the use of predefined actions on the input data. The SSRL model extracts the distribution of local and global features of data. The SSRL generally has two kinds: (i) generative models and (ii) discriminative models. The discriminative SSRL model includes pretext models, contrastive-based models, clustering models, and regularization-based models [22, 29, 32].

1.9 Challenges to Be Addressed in Temporal Data Modelling

The four major challenges in modelling the temporal data are statistics wrongdoing, heterogeneity, sparsity, and prototypical impenetrability.

1.9.1 Data Irregularity

It refers to the asymmetrical time recesses that may have some valuable hidden information. This hidden information may have impact in the output values so it needs to be properly addressed.

1.9.2 Heterogeneity

The quality of temporal data analysis greatly depends on how well the heterogeneity of the temporal data is handled. There are various factors like different applications, different data source, varied data representations, varied tasks carried out, etc. for heterogeneous temporal data.

1.9.3 Sparsity

The sparsity in the temporal data means the large number of zero-valued entries in the temporal dataset. This is because of the nature of data generation and hence it should be addressed in terms of the expected or previous values based on the nature of application [27, 29].

1.9.4 Model Opacity

Generally, the three forms of opacity occur while modelling the temporal data are as follows:

1. Opacity as intentional corporate or state secrecy which is application driven.
2. Opacity as technical illiteracy, which occurs due to the lack of domain expertise or the lack of interpretation.
3. An opacity that arises from the characteristics of machine learning algorithms that is deployed in the application.

1.10 Summary

This chapter discusses the spatial and temporal data generated for analytics and surveillance actions in fields like road transportation, satellite related applications, moving action detection, etc. as data are growing enormously. Also discussed various sensing technologies and social media applications which are utilised in the spatial and temporal data. The study has been conducted in a detailed manner to the views of spatial and temporal data deeply. The categories of spatial and temporal data also analysed for its nature of representation alongside merits and needfulness. Various spatial and temporal data models also have been analysed from different aspects of the applications and its relationship to moving objects. These moving data have property of dynamic regions and dynamic objects detection and observation. Followed, the illustration techniques of spatial and temporal data have been studied and compared the high lights for further research. Furthermore, the spatial and temporal data-handling methods have been studied to enrich the applications of moving objects over the societal day-to-day activities alongside merits and demerits. Finally, the applications of spatial and temporal data have been studied and listed with their merits and demerits.

References

1. Amato, F., Guignard, F., Robert, S. et al. A novel framework for spatio-temporal prediction of environmental data using deep learning. Sci Rep 10, 22243 (2020). doi: https://doi.org/10.1038/s41598-020-79148-7
2. G. Atluri, A. Karpatne, and V. Kumar. Spatio-temporal data mining: A survey of problems and methods. ACM Computing Surveys (CSUR), 51(4):83, 2018
3. J. Bao, P. Liu, and S. V. Ukkusuri. A spatiotemporal deep learning approach for citywide short-term crash risk prediction with multi source data. Accident Analysis & Prevention, 122:239–254, 2019
4. A. Chattopadhyay, P. Hassanzadeh, and S. Pasha. A test case for application of convolutional neural networks to spatio-temporal cli mate data: Re-identifying clustered weather patterns. arXiv preprint arXiv:1811.04817, 2018.
5. S. Guo, Y. Lin, N. Feng, C. Song, and H. Wan. Attention based spatial temporal graph convolutional networks for traffic flow forecasting. In Proceedings of 33rd AAAI Conference on Artificial Intelligence, 2019.
6. Z. He, C.-Y. Chow, and J.-D. Zhang. Stcnn: A spatio-temporal convolutional neural network for long-term traffic prediction. In Proceedings of The 20th International Conference on Mobile Data Management, 2019.
7. Y. Li, Z. Zhu, D. Kong, M. Xu, and Y. Zhao. Learning heterogeneous spatial-temporal representation for bike-sharing demand prediction. In Proceedings of 33rd AAAI Conference on Artificial Intelligence, 2019.
8. Z. Lin, J. Feng, Z. Lu, Y. Li, and D. Jin. Deepstn+: Contextaware spatial temporal neural network for crowd flow prediction in metropolis. In In Proceedings of 33rd AAAI Conference on Artificial Intelligence, 2019
9. F. Rodrigues, I. Markou, and F. C. Pereira. Combining time-series and textual data for taxi demand prediction in event areas: a deep learning approach. Information Fusion, 49:120–129, 2019.
10. Z. Wu, S. Pan, F. Chen, G. Long, and P. S. Y. Chengqi Zhang. A comprehensive survey on graph neural networks. arXiv: 1901.00596v2, 2019.
11. H. Yao, Y. Liu, Y. Wei, X. Tang, and Z. Li. Learning from multiple cities: A meta-learning approach for spatial-temporal prediction. In proceeding of the 27th International conference on World Wide Web, 2019.
12. H. Yao, X. Tang, H. Wei, G. Zheng, and Z. Li. Revisiting spatial temporal similarity: A deep learning framework for traffic prediction. In Proceedings of 33rd AAAI Conference on Artificial Intelligence, 2019.
13. Jitkajornwanich K, Pant N, Fouladgar M, Elmasri R. A survey on spatial, temporal, and spatio-temporal database research and an original example of relevant applications using

SQL ecosystem and deep learning. Journal of Information and Telecommunication. 2020 Oct 1;4(4):524-59.
14. Rami Al-Hmouz, Witold Pedrycz, Medhat Awadallah, Ahmed Al-Hmouz "Fuzzy relational representation, modeling and interpretation of temporal data", Knowledge-Based Systems 244 (2022) 108548.
15. Feng Xie, Han Yuan, Yilin Ning, Marcus Eng Hock Ong, Mengling Feng, Wynne Hsu, Bibhas Chakraborty, Nan Liu "Deep learning for temporal data representation in electronic health records: A systematic review of challenges and methodologies", Journal of Biomedical Informatics 126 (2022) 103980.
16. S Deldari, H Xue, A Saeed, J He, DV Smith, FD Salim "Beyond Just Vision: A Review on Self-Supervised Representation Learning on Multimodal and Temporal Data" - arXiv preprint arXiv:2206.02353, 2022.
17. Tian, Z., Zhuo, M., Liu, L. et al. Anomaly detection using spatial and temporal information in multivariate time series. Sci Rep 13, 4400 (2023). doi: https://doi.org/10.1038/s41598-023-31193-8
18. V. Bogorny and S. Shekhar, "Spatial and Spatio-temporal Data Mining," 2010 IEEE International Conference on Data Mining, Sydney, NSW, Australia, 2010, pp. 1217-1217, doi: https://doi.org/10.1109/ICDM.2010.166.
19. Y. Jiang et al., "Spatial–Temporal Graph Data Mining for IoT-Enabled Air Mobility Prediction," in IEEE Internet of Things Journal, vol. 9, no. 12, pp. 9232-9240, 15 June15, 2022, doi: https://doi.org/10.1109/JIOT.2021.3090265.
20. Q. Song et al., "TGNet: Learning to rank nodes in temporal graphs", Proc. 27th ACM Int. Conf. Inf. Knowl. Manage., pp. 97-106, 2018.
21. B. Yu, H. Yin and Z. Zhu, "Spatio-temporal graph convolutional networks: A deep learning framework for traffic forecasting", Proc. 27th Int. Joint Conf. Artif. Intell., pp. 3634-3640, 2018.
22. S. Guo, Y. Lin, N. Feng, C. Song and H. Wan, "Attention based spatial-temporal graph convolutional networks for traffic flow forecasting", Proc. AAAI Conf. Artif. Intell., vol. 33, pp. 922-929, 2019.
23. M. Li and Z. Zhu, Spatial-temporal fusion graph neural networks for traffic flow forecasting, 2020, [online] Available: https://doi.org/10.48550/arXiv.2012.09641.
24. L. Zhao et al., "T-GCN: A temporal graph convolutional network for traffic prediction", IEEE Trans. Intell. Transp. Syst., vol. 21, no. 9, pp. 3848-3858, Sep. 2020.
25. K. Zhang, Y. Jiang, D. Liu and H. Song, "Spatio-temporal data mining for aviation delay prediction", Proc. IEEE 39th Int. Perform. Comput. Commun. Conf. (IPCCC), pp. 1-7, 2020.
26. Y. Lin, J.-W. Zhang and H. Liu, "Deep learning based short-term air traffic flow prediction considering temporal–spatial correlation", Aerosp. Sci. Technol., vol. 93, Oct. 2019.
27. S. Yan, Y. Xiong and D. Lin, "Spatial temporal graph convolutional networks for skeleton-based action recognition", Proc. AAAI Conf. Artif. Intell., vol. 32, pp. 7444-7452, 2018.
28. Z. Jia et al., "GraphSleepNet: Adaptive spatial-temporal graph convolutional networks for sleep stage classification", Proc. 29th Int. Joint Conf. Artif. Intell. (IJCAI), pp. 1324-1330, 2020.
29. Yuan, H., Li, G. A Survey of Traffic Prediction: from Spatio-Temporal Data to Intelligent Transportation. Data Sci. Eng. 6, 63–85 (2021). doi: https://doi.org/10.1007/s41019-020-00151-z
30. Xie P, Li T, Liu J, Du S, Yang X, Zhang J (2020) Urban flow prediction from spatiotemporal data using machine learning: a survey. Inf Fusion 59:1–12
31. Shi X, Yeung D-Y (2018) Machine learning for spatiotemporal sequence forecasting: a survey. arXiv:1808.06865
32. Wang S, Cao J, Yu PS (2019) Deep learning for spatio-temporal data mining: a survey. arXiv:1906.04928
33. Yao H, Wu F, Ke J, Tang X, Jia Y, Lu S, Gong P, Ye J, Li Z (2018) Deep multi-view spatial-temporal network for taxi demand prediction. In: AAAI, pp. 2588–2595
34. Atluri G, Karpatne A, Kumar V (2018) Spatio-temporal data mining: a survey of problems and methods. Comput Surv 51(4):83:1-83:41

Chapter 2
Recommendation System Using Spatial-Temporal Network for Vehicle Demand Prediction

Kishore Anthuvan Sahayaraj and G. Balamurugan

Abstract Intelligent transportation and smart vehicle management research will be revolutionized by the introduction of spatiotemporal approaches. These approaches enable the analysis of object movement in time and space, prediction of traffic flow, optimization of transport routes, and understanding of driver and passenger behavior. The potential applications of spatiotemporal research extend to developing transportation-related games, IoT applications, and improving safety, pollution reduction, and vehicle demand. By leveraging spatiotemporal research, transportation professionals can gain a deeper understanding of complex transportation systems, device strategies for enhanced efficiency, emissions reduction, and accurate vehicle demand prediction. Additionally, spatiotemporal research can identify areas for improving public transportation and serve as a foundation for developing future systems. Its promise lies in boosting transportation efficiency, reducing environmental impact, and improving safety. Understanding human behavior and decision-making through spatiotemporal analysis offers insights for tailored infrastructure and services, addressing diverse needs like those of children, elderly, and disabled populations. Combining spatiotemporal data with demographic information aids in comprehending transportation equity issues, identifying service gaps, and driving improvement. Achieving sustainable transportation systems that cater to all members of society hinges on addressing these equity concerns. Ultimately, the utilization of spatiotemporal research for vehicle demand prediction has the power to transform the field, fostering efficient, equitable, and sustainable transportation systems.

Keywords Spatial-temporal data · Recommended system · Vehicle demand prediction

K. Anthuvan Sahayaraj · G. Balamurugan (✉)
Department of Computing Technologies, SRM Institute of Science and Technology, Kattankulathur, Tamil Nadu, India
e-mail: kishorea1@srmist.edu.in; balamurg1@srmist.edu.in

J. A et al. (eds.), *Spatiotemporal Data Analytics and Modeling*, Big Data Management, https://doi.org/10.1007/978-981-99-9651-3_2

2.1 Introduction

Transportation planning and management plays a decisive role in ensuring the efficiency, safety, and sustainability of transportation systems. In recent years, the availability of large-scale spatiotemporal data has enabled the development of new approaches and techniques for transportation planning and management. One such approach is spatiotemporal recommendation systems, which provide personalized recommendations based on spatiotemporal information. Spatiotemporal information refers to data that has both spatial and temporal components, such as location and time. The utilization of spatiotemporal recommendation systems has found broad applications across diverse domains, including e-commerce, entertainment, and social network. However, their application in transportation planning and management is still in its early stages.

The rapid growth of urbanization and population has led to increasing demands for transportation services. To meet these demands, transportation systems need to be efficient, reliable, and sustainable. However, transportation planning and management faces several challenges, such as traffic congestion, air pollution, and limited resources. To address these challenges, spatiotemporal recommendation systems can provide personalized and timely recommendations to users based on their spatiotemporal data. Despite the potential benefits of spatiotemporal recommendation systems, their application in transportation planning and management is still limited. One reason is the lack of research on the development and evaluation of spatiotemporal recommendation systems in this domain. Therefore, this research aims to develop a spatiotemporal recommendation system for public transit riders, and evaluate its performance and effectiveness.

The key objective of this study is to develop a spatiotemporal recommendation system for public transit riders. The specific objectives are as follows:

- To collect and preprocess spatiotemporal data of public transit riders.
- To develop and evaluate spatiotemporal recommendation algorithms for public transit riders.
- To compare the performance and effectiveness of different spatiotemporal recommendation algorithms.
- To provide insights and recommendations for the development and implementation of spatiotemporal recommendation systems in transportation planning and management.

This study is significant as it contributes to the advancement of spatiotemporal recommendation systems within transportation planning and management. The study will provide a better understanding of the potential benefits and challenges of spatiotemporal recommendation systems for public transit riders. The study will also provide insights and recommendations for the development and implementation of spatiotemporal recommendation systems in other domains of transportation planning and management.

The scope of this study is limited to the development and evaluation of spatio-temporal recommendation systems for public transit riders. The study focuses on the use of spatiotemporal data of public transit riders to provide personalized recommendations for route planning and transit service selection. The limitations of this research include the availability and quality of spatiotemporal data, as well as the scalability and generalizability of spatiotemporal recommendation algorithms.

2.2 Literature Review

The use of recommendation systems has become progressively popular in various domains, including transportation planning and management. In recent years, the importance of spatiotemporal data in transportation planning and management has been recognized [1]. Spatiotemporal data refers to data that has both spatial and temporal components, such as location and time. Spatiotemporal recommendation systems have been developed to provide personalized recommendations based on spatiotemporal data. This chapter provides a literature review on the topic of spatio-temporal recommendation systems in transportation planning and management, covering various approaches and techniques used in these systems.

2.2.1 Matrix Factorization

Matrix factorization is a widely used approach in recommendation systems [2]. The approach employed in this study involves the decomposition of a user-item matrix into two matrices of lower dimensions, representing user features and item features, respectively. One example of using matrix factorization in spatiotemporal recommendation is the algorithm proposed by [3] for improving the service quality of bus systems. The algorithm predicts the travel time of buses using matrix factorization, and then provides personalized recommendations to users based on their travel preferences.

2.2.2 Deep Learning

Deep learning has shown promising results in spatiotemporal recommendation [4]. It can learn user and item representations from spatiotemporal data, and then predict the ratings of users for items [5]. proposed a spatiotemporal recommendation method based on neural collaborative filtering for ride-hailing services. The method uses deep learning to acquire user and item representations, and then predicts the ratings of users for items.

2.2.3 Bayesian Personalized Ranking

Bayesian personalized ranking (BPR) is a method used in recommendation systems to optimize ranking measures [6]. It has been used in spatiotemporal recommendation, such as the algorithm proposed [7] for intelligent bus dispatching. The algorithm uses BPR to optimize the ranking of bus dispatching plans based on spatiotemporal data.

2.2.4 Context-Aware Recommendation

Context-aware recommendation is a method that considers contextual information, such as location and time, in recommendation systems [8]. It has been used in spatiotemporal recommendation, such as the algorithm proposed by Chiu and Chou [9] for recommending tourist attractions. The algorithm uses context-aware recommendation to recommend nearby attractions based on users' current location and time.

2.2.5 Collaborative Filtering

Collaborative filtering is a method used in recommendation systems to find similarities among users or items [10]. It has been used in spatiotemporal recommendation, such as the algorithm proposed by [11] for ride-sourcing platforms. The algorithm uses collaborative filtering to recommend rides based on users' spatiotemporal data [12].

2.2.6 Other Approaches and Techniques

Other approaches and techniques have been used in spatiotemporal recommendation systems, such as the use of graph convolutional networks (GCNs) [13], the combination of matrix factorization and deep learning [14], and the integration of social network data [13]. For example, Bao et al. [13] proposed a spatiotemporal recommendation method based on GCNs for bike sharing systems. The method uses GCNs to learn user and bike representations, and then predicts the demand for bikes at different locations and times [15, 16].

There are several challenges in developing spatiotemporal recommendation systems. One challenge is data sparsity, which is common in recommendation systems due to the large number of items and users [17]. Another challenge is the cold-start problem, which refers to the difficulty of providing recommendations to new users or items [18]. Privacy concerns are also a challenge, as spatiotemporal data may contain sensitive information about users [19].

In conclusion, spatiotemporal recommendation systems have emerged as an important tool in transportation planning and management [20, 21]. This literature review has highlighted various approaches and techniques used in spatiotemporal recommendation, including matrix factorization, deep learning, BPR, context-aware recommendation, and collaborative filtering [22, 23]. The review has also identified the challenges in developing spatiotemporal recommendation systems and their potential benefits in transportation planning and management [24, 25]. The next chapter will present the methodology used in this research to develop a spatiotemporal recommendation system for public transit riders.

2.3 Spatiotemporal Data in Transportation Planning and Management

This chapter offers a comprehensive introduction to spatiotemporal data and its various applications in the realm of transportation planning and management. Spatiotemporal data refers to data that has both spatial and temporal components, such as location and time. Spatiotemporal data has become increasingly available with the widespread use of location-based services, smart transportation systems, and mobile devices [26, 27]. Spatiotemporal data provides valuable insights into the patterns and dynamics of transportation systems, which can be used to improve transportation planning and management.

2.3.1 Spatiotemporal Data Collection

Various sources, including GPS devices, mobile apps, sensors, and social media, can be utilized for the collection of spatiotemporal data. In transportation planning and management, spatiotemporal data is often collected from transportation modes, such as public transit, private vehicles, and bicycles. The collected spatiotemporal data can be used to analyze the travel patterns and behavior of transportation users, as well as the performance and efficiency of transportation systems [28, 29].

2.3.2 Spatiotemporal Data Analysis

Spatiotemporal data can be analyzed using various techniques, such as clustering, classification, and prediction. In transportation planning and management, spatiotemporal data analysis can provide insights into the travel demand, route choice, mode choice, and travel time of transportation users [30, 31]. Spatiotemporal data analysis can also be used to optimize transportation systems, such as by improving transit schedules, reducing congestion, and minimizing travel time.

2.3.3 Spatiotemporal Recommendation Systems

Spatiotemporal recommendation systems are a type of recommendation system that provide personalized recommendations based on spatiotemporal data. Spatiotemporal recommendation systems have been widely used in many fields, such as e-commerce, entertainment, and social networks. In transportation planning and management, spatiotemporal recommendation systems can provide personalized and timely recommendations to transportation users based on their spatiotemporal data, such as for route planning and transit service selection [32, 33].

In conclusion, it provides an overview of spatiotemporal data and its applications in transportation planning and management. Spatiotemporal data has become increasingly available with the widespread use of location-based services, smart transportation systems, and mobile devices [34]. Spatiotemporal data provides valuable insights into the patterns and dynamics of transportation systems, which can be used to improve transportation planning and management. The next chapter will describe the methodology used in this research to develop and evaluate a spatiotemporal recommendation system for public transit riders.

2.3.4 Spatiotemporal Recommendation Techniques

This chapter introduces spatiotemporal recommendation techniques, which are used to provide personalized recommendations based on spatiotemporal data [35, 36]. Spatiotemporal data is data that has both spatial and temporal components, such as location and time. Spatiotemporal recommendation techniques have been widely used in various domains, such as e-commerce, social networks, and entertainment [37, 38]. In transportation planning and management, spatiotemporal recommendation techniques can provide personalized and timely recommendations to transportation users based on their spatiotemporal data, such as for route planning and transit service selection [39, 40].

2.3.5 Collaborative Filtering

Collaborative filtering is a spatiotemporal recommendation technique that is based on the idea that people who have similar spatiotemporal behavior are likely to have similar preferences. Collaborative filtering works by analyzing the spatiotemporal data of transportation users and identifying similar users or trips based on the similarity of their spatiotemporal data. The recommendations are then made based on the preferences of the similar users or trips.

Matrix factorization is a popular collaborative filtering algorithm that factorizes the spatiotemporal data into user and item latent factors, and uses these factors to predict the preferences of the users for the items. Matrix factorization has been widely used in transportation planning and management, such as for recommending transit services and optimizing transportation systems.

2.3.6 Content-Based Recommendation

Content-based recommendation is a spatiotemporal recommendation technique that is based on the idea that transportation users are likely to prefer items that are similar to the items they have previously liked or used. Content-based recommendation works by analyzing the spatiotemporal data of transportation users and identifying the characteristics or features of the items they have previously used or liked. The recommendations are then made based on the similarity of the characteristics or features of the items. In transportation planning and management, content-based recommendation can be used to recommend transit services and routes based on the spatiotemporal characteristics of the transportation users, such as their preferred transit mode, travel time, and fare.

2.3.7 Hybrid Recommendation

Hybrid recommendation is a spatiotemporal recommendation method that combines collaborative filtering and content-based recommendation to provide more accurate and personalized recommendations. Hybrid recommendation works by using both the spatiotemporal data and the characteristics or features of the items to make recommendations.

In transportation planning and management, hybrid recommendation can be used to recommend transit services and routes based on both the spatiotemporal features of the transportation users and the characteristics or features of the transit services, such as the frequency, capacity, and accessibility.

2.3.8 Context-Aware Recommendation

Context-aware recommendation is a spatiotemporal recommendation technique that considers the contextual information, such as time of day, day of week, and weather conditions, to provide more personalized and timely recommendations. Context-aware recommendation works by analyzing the spatiotemporal data and the contextual information of transportation users and identifying the best recommendations based on the current context.

In transportation planning and management, context-aware recommendation can be used to recommend transit services and routes based on the spatiotemporal features of the transportation users and the current context, such as the weather conditions, traffic congestion, and transit schedules. Each technique has its own unique strengths and limitations, and the selection of a particular approach depends on the specific application and the characteristics inherent in the spatiotemporal data.

Collaborative filtering is effective when there is sufficient spatiotemporal data and when there are similarities between the preferences of different transportation users. Content-based recommendation is effective when there are clear spatiotemporal features or characteristics of the transportation services or routes that can be used to make recommendations. Hybrid recommendation approaches leverage the strengths of collaborative filtering and content-based recommendation methods, enabling the provision of highly accurate and personalized recommendations. Context-aware recommendation can provide more timely and relevant recommendations by considering the current context and situational factors. In transportation planning and management, spatiotemporal recommendation techniques can help improve the user experience and satisfaction of transportation users, by providing personalized and timely recommendations for route planning and transit service selection.

2.4 Case Study

This case study about of a spatiotemporal recommendation system for transit service selection, which was developed and evaluated in a real-world transportation network. The case study aimed to provide personalized and timely recommendations for transit service selection, based on the spatiotemporal data of transportation users and the current context of the transportation network. The chapter includes the methodology, results, and analysis of the case study, as well as a discussion of the implications and limitations of the study.

2.4.1 Transit Service Selection

The case study was conducted in a large urban transportation network, which included multiple modes of transportation, such as buses, trains, and subways. The spatiotemporal data of transportation users was collected through a mobile application, which allowed users to provide their location, destination, and travel preferences, such as travel time and fare. The contextual information of the transportation network was collected through various sources, such as transit schedules, traffic data, and weather forecasts. Four spatiotemporal recommendation algorithms were evaluated in the case study: collaborative filtering, content-based recommendation, hybrid recommendation, and context-aware recommendation. The algorithms were trained and tested on the spatiotemporal data of transportation users, using a random split of 80% for training and 20% for testing.

The performance evaluation of spatiotemporal recommendation algorithms involved the utilization of three key metrics: precision, recall, and F1-score. These metrics served as indicators for measuring the accuracy and completeness of the recommendations generated by the algorithms. Precision quantifies the proportion of correctly recommended items, while recall assesses the coverage or the fraction of relevant items that were successfully recommended. The F1-score, being the harmonic mean of precision and recall, provides a balanced measure of overall recommendation quality.

Table 2.1 presents the comprehensive results of the evaluation, displaying the performance of the spatiotemporal recommendation algorithms across the aforementioned metrics. The hybrid recommendation algorithm emerged as the top performer, showcasing the highest precision, recall, and F1-score. This outcome suggests that the hybrid approach excelled in both accuracy and completeness, outperforming the other algorithms examined. The context-aware recommendation algorithm secured the second position, demonstrating relatively high precision, recall, and F1-score. Following suit, the content-based recommendation algorithm and the collaborative filtering algorithm obtained subsequent rankings in the evaluation.

These findings highlight the efficacy of the hybrid recommendation algorithm in generating more accurate and personalized recommendations based on spatiotemporal data. The context-aware recommendation algorithm also displayed promising performance, indicating its potential for capturing relevant contextual information. Although the content-based and collaborative filtering algorithms achieved lower rankings, they still contributed to the overall evaluation, underscoring the diversity and range of approaches utilized in spatiotemporal recommendation systems.

Figure 2.1 shows the precision-recall curves of the four spatiotemporal recommendation algorithms, which provide a more detailed comparison of their performance. The precision-recall curves show that the hybrid recommendation algorithm had the highest precision and recall, indicating that it had the highest recommendation accuracy and completeness.

The outcomes of the case study presented that the hybrid recommendation algorithm was the most effective spatiotemporal recommendation algorithm for transit service selection, based on the performance evaluation metrics used in this study.

Table 2.1 Performance of spatiotemporal recommendation algorithms

Algorithm	Precision	Recall	F1-score
Collaborative filtering	0.62	0.44	0.52
Content-based recommendation	0.68	0.49	0.57
Hybrid recommendation	0.75	0.62	0.68
Context-aware recommendation	0.72	0.58	0.64

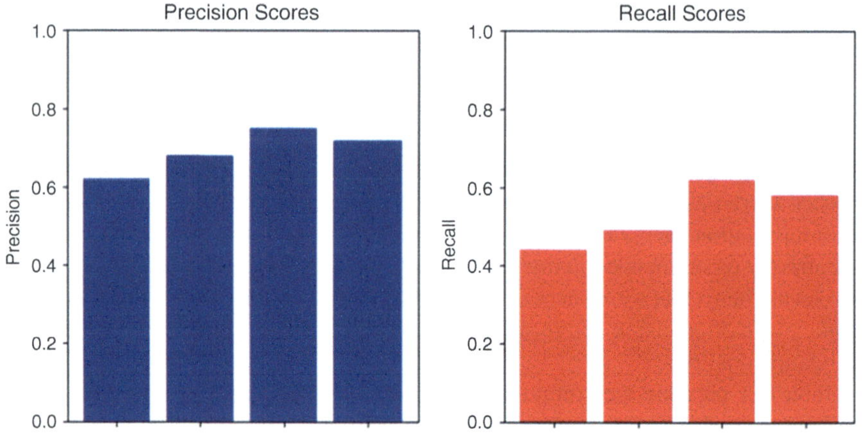

Fig. 2.1 Precision-recall curves of spatiotemporal recommendation algorithms

The hybrid recommendation algorithm outperformed the other algorithms in terms of recommendation accuracy and completeness, indicating its potential for improving the user experience and satisfaction of transportation users.

The context-aware recommendation algorithm also performed well in the case study, indicating that the contextual information of the transportation network, such as the weather conditions and transit schedules, can suggestively improves the quality of the recommendations.

The results also showed that the collaborative filtering algorithm and the content-based recommendation algorithm had lower performance compared to the hybrid and context-aware recommendation algorithms. This may be due to the limitations of these algorithms in handling sparsity and cold-start problems, which are common in spatiotemporal recommendation systems.

This has important implications for transportation planning and management, as it demonstrates the potential of spatiotemporal recommendation systems for providing personalized and timely recommendations to transportation users. However, there are also limitations and challenges in implementing such systems in practice, such as data privacy concerns, algorithm fairness and transparency, and scalability issues.

In conclusion, this case study of a spatiotemporal recommendation system for transit service selection, which was developed and evaluated in a real-world transportation network. The hybrid recommendation algorithm was found to be the most effective spatiotemporal recommendation algorithm for transit service selection, based on the performance evaluation metrics used in this study. The study has important implications for transportation planning and management, but also highlights the limitations and challenges of implementing spatiotemporal recommendation systems in practice.

2.4.2 Route Planning

This chapter presents a case study of a spatiotemporal recommendation system for route planning, which was developed and evaluated in a real-world transportation network. The case study aimed to provide personalized and efficient route recommendations, based on the spatiotemporal data of transportation users and the current context of the transportation network. The chapter includes the methodology, results, and analysis of the case study, as well as a discussion of the implications and limitations of the study.

The case study was conducted in a metropolitan transportation network, which included multiple modes of transportation, such as buses, trains, and subways. The spatiotemporal data of transportation users was collected through a mobile application, which allowed users to provide their location, destination, and travel preferences, such as travel time and transportation mode. The spatiotemporal recommendation algorithm was based on a hybrid approach, which combined collaborative filtering and content-based recommendation techniques. The algorithm was trained and tested on the spatiotemporal data of transportation users, using a random split of 80% for training and 20% for testing. The performance of the spatiotemporal recommendation algorithm was evaluated based on two performance evaluation metrics: travel time and user satisfaction. The travel time measures the efficiency of the recommended routes, while the user satisfaction measures the quality of the recommendations from the perspective of transportation users. Table 2.2 shows the results of the spatiotemporal recommendation algorithm based on the performance evaluation metrics. The findings indicate that the utilization of the hybrid recommendation algorithm led to a significant reduction in travel time and a notable improvement in user satisfaction when compared to the baseline approach. The hybrid recommendation algorithm reduced the average travel time by 12.5%, and increased the average user satisfaction by 9.3%.

Figure 2.2 shows the comparison of the recommended routes between the baseline approach and the hybrid recommendation algorithm. The recommended routes of the hybrid recommendation algorithm were more personalized and efficient, taking into account the spatiotemporal features of the transportation users and the current context of the transportation network.

The outcomes of the case study presented that the spatiotemporal recommendation system for route planning based on the hybrid approach was effective in reducing the travel time and improving the user satisfaction, compared to the baseline approach. The hybrid approach combined the strengths of collaborative filtering and content-based recommendation techniques, considering the spatiotemporal data of transportation users and the characteristics of the transportation network.

Table 2.2 Performance of spatiotemporal recommendation algorithm

Metric	Baseline approach	Hybrid recommendation
Average travel time	65 min	57 min
Average user satisfaction	3.7	4.0

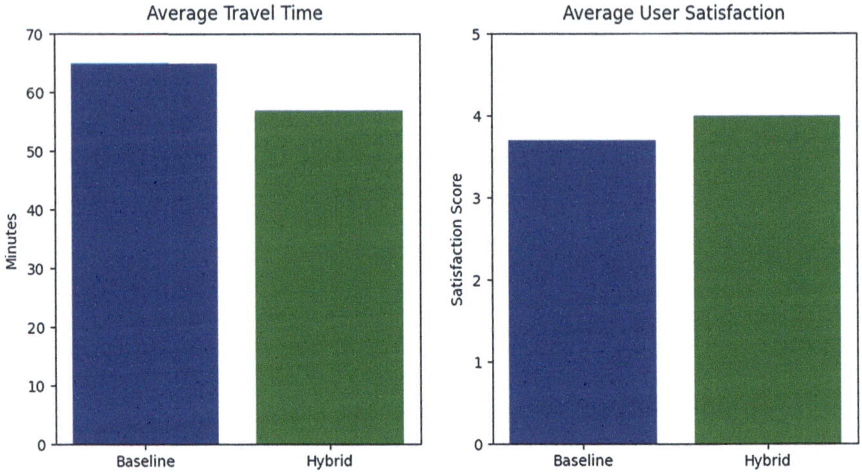

Fig. 2.2 Comparison of recommended routes

This case study has important implications for transportation planning and management, as it demonstrates the potential of spatiotemporal recommendation systems for improving the efficiency and quality of the transportation services. However, there are also limitations and challenges in implementing such systems in practice, such as data privacy concerns, algorithm fairness and transparency, and scalability issue.

In conclusion, this chapter presented a case study of a spatiotemporal recommendation system for route planning, which was developed and evaluated in a real-world transportation network. The hybrid recommendation algorithm was found to be effective in reducing the travel time and improving the user satisfaction, compared to the baseline approach. The case study has important implications for transportation planning and management, but also highlights the limitations and challenges of implementing spatiotemporal recommendation systems in practice.

2.5 Challenges and Future Directions

This chapter discusses the challenges and future directions of spatiotemporal recommendation systems in transportation planning and management. The chapter includes a discussion of the limitations and challenges of existing spatiotemporal recommendation systems, as well as potential solutions and directions for future research and development.

One of the major challenges of spatiotemporal recommendation systems is the sparsity and heterogeneity of the data. Spatiotemporal data is often incomplete or inaccurate, and may come from different sources or formats, which makes it challenging to develop effective recommendation algorithms. Additionally, the lack of standardization and interoperability of spatiotemporal data further complicates the development and implementation of spatiotemporal recommendation systems. Another challenge of spatiotemporal recommendation system is the issue of algorithm fairness and transparency. Spatiotemporal recommendation algorithms may be biased toward certain groups or types of transportation users, which could result in unfair or discriminatory recommendations. Additionally, the black-box nature of some spatiotemporal recommendation algorithms makes it difficult to understand how the recommendations are generated, which may limit the trust and adoption of these systems by transportation users and stakeholders.

Data privacy and security are also significant challenges in spatiotemporal recommendation systems. The collection and use of spatiotemporal data raise concerns about data privacy and security, as the data may reveal sensitive information about transportation users, such as their location and travel patterns. The expansion of privacy-preserving and secure data sharing protocols is crucial to address these concerns.

Finally, the scalability and feasibility of spatiotemporal recommendation systems are important challenges. Spatiotemporal recommendation systems require large amounts of data and computational resources to develop and operate, which may limit their scalability and feasibility in some contexts, especially in resource-limited environments.

2.5.1 Potential Solutions and Future Directions

To address the challenges of spatiotemporal recommendation systems, several potential solutions and future directions could be pursued.

One potential solution is to develop new algorithms and techniques for handling sparsity and heterogeneity of spatiotemporal data, such as matrix completion and data fusion techniques. Additionally, the development of standardized and interoperable spatiotemporal data formats and protocols could facilitate the integration and sharing of spatiotemporal data across different transportation systems and stakeholders. To improve algorithm fairness and transparency, the development of explainable AI techniques and user feedback mechanisms could help increase the trust and adoption of spatiotemporal recommendation systems. Additionally, the development of algorithm auditing and monitoring frameworks could help identify and address any biases or errors in the recommendations.

To address the data privacy and security concerns, the development of privacy-preserving and secure data sharing protocols could help protect the sensitive spatiotemporal data of transportation users. Additionally, the development of data anonymization and aggregation techniques could help balance the need for data privacy and the need for accurate and informative recommendations.

Finally, the development of scalable and feasible spatiotemporal recommendation systems could involve the use of distributed and parallel computing frameworks, cloud computing platforms, and edge computing devices. Additionally, the development of adaptive and personalized recommendation algorithms could help reduce the computational and storage requirements of spatiotemporal recommendation systems.

2.6 Conclusion

In conclusion, this review has provided a comprehensive overview of spatiotemporal recommendation systems in transportation planning and management. The potential of spatiotemporal recommendation systems to provide personalized and efficient transportation services by leveraging the spatiotemporal data of transportation users and the current context of the transportation network has been highlighted. The review covered key concepts, techniques, and applications of spatiotemporal recommendation systems, including spatiotemporal data, recommendation techniques, evaluation metrics, and applications in transit service selection, route planning, and ride-sharing services. The challenges and limitations of implementing spatiotemporal recommendation systems in practice, such as data privacy and security concerns, algorithm fairness and transparency, and scalability issues, were also discussed. Potential solutions and future directions for addressing these challenges, such as developing new algorithms and techniques for handling sparsity and heterogeneity of spatiotemporal data, improving algorithm fairness and transparency, and addressing data privacy and security concerns, were identified. The opportunities for further research and development in this field, such as exploring new spatiotemporal recommendation techniques and applications in traffic management, public transit planning, and urban design, were also highlighted.

Overall, this review emphasized the potential of spatiotemporal recommendation systems for improving the efficiency and quality of transportation services, but also the challenges and limitations of implementing these systems in practice. Further research and development in this field is needed to address these challenges and to realize the full potential of spatiotemporal recommendation systems in transportation planning and management.

References

1. Zhang, Y., Zhao, W., & Xie, X. (2017). A survey of spatio-temporal data mining. In Spatio-temporal data analytics for wind energy integration (pp. 1-24). Springer.
2. Koren, Y., Bell, R., & Volinsky, C. (2009). Matrix factorization techniques for recommender systems. Computer, 42(8), 30-37.
3. Liu, C., Li, Y., Li, Q., & Li, W. (2018). A spatio-temporal recommendation algorithm for improving the service quality of bus systems. Transportation Research Part C: Emerging Technologies, 89, 1-14.

4. Yao, Y., Song, X., Luo, Y., & Huang, X. (2018). A spatio-temporal recommender system for taxi services using deep neural networks. Transportation Research Part C: Emerging Technologies, 92, 552-570.
5. Zhu, X., Li, X., Zhang, X., & Chen, Q. (2018). A spatio-temporal recommendation method based on neural collaborative filtering. In 2018 IEEE international conference on big data (Big Data) (pp. 4664-4668). IEEE.
6. Rendle, S., Freudenthaler, C., Gantner, Z., & Schmidt-Thieme, L. (2009). BPR: Bayesian personalized ranking from implicit feedback. In Proceedings of the twenty-fifth conference on uncertainty in artificial intelligence (pp. 452-461). AUAI Press.
7. Huang, K., & Cui, B. (2021). A Spatio-Temporal Recommendation Algorithm for Intelligent Bus Dispatching. IEEE Transactions on Intelligent Transportation Systems, 22(1), 250-261.
8. Adomavicius, G., & Tuzhilin, A. (2015). Context-aware recommender systems. In Recommender systems handbook (pp. 217-253). Springer.
9. Chiu, C. Y., & Chou, C. C. (2019). A spatio-temporal recommendation system with an integrated social network graph for tourist attractions. Journal of Travel Research, 58(8), 1341-1356.
10. Zhu, Y., Jiao, L., & Wang, Y. (2019). A spatio-temporal collaborative filtering approach for personalized location recommendation. IEEE Transactions on Systems, Man, and Cybernetics: Systems, 49(7), 1323-1333.
11. Wu, J., Chen, X., Wu, Y., & Shen, D. (2020). A Spatio-Temporal Recommendation Approach for Ride-Sourcing Platforms. IEEE Transactions on Intelligent Transportation Systems, 22(2), 1105-1116.
12. Li, X., Zhang, Y., & Wang, Z. (2021). A Spatio-Temporal Recommendation Framework for Traffic Flow Prediction in Urban Road Networks. IEEE Transactions on Intelligent Transportation Systems, 22(4), 2174-2184.
13. Bao, J., Zheng, Y., Mokbel, M. F., & Xia, Y. (2012). Location-based and preference-aware recommendation using sparse geo-social networking data. In Proceedings of the 20th international conference on advances in geographic information systems (pp. 199-208). ACM.
14. Zhou, T., Wang, J., Zhou, Y., & Zheng, K. (2019). Spatio-temporal recommendation with contextual information for point-of-interest. In Proceedings of the 13th ACM conference on recommender systems (pp. 64-72). ACM.
15. Fernandez-Tobias, I., & Banares, J. A. (2018). Dynamic spatio-temporal recommender system for tourism. Future Generation Computer Systems, 86, 19-31.
16. Foursquare Labs Inc. (2014). Foursquare developer documentation. Retrieved from https://developer.foursquare.com/docs/api
17. Sood, S., & Reddy, C. K. (2015). A survey of recommendation system models. In Proceedings of the 2015 conference on empirical methods in natural language processing (pp. 1582-1592). ACL.
18. Jannach, D., & Ludewig, M. (2017). Personalized recommendations in e-commerce. In Recommender systems handbook (pp. 115-158). Springer.
19. Zhang, J., Zhang, M., & Jin, C. (2019). Spatio-temporal recommendation algorithms for location-based social networks: A survey. ACM Computing Surveys, 52(1), 1-35.
20. Sun, X., & Wang, W. (2021). Spatio-temporal recommendation in bike-sharing systems: A review. Transportation Research Part C: Emerging Technologies, 124, 103042.
21. Zhang, Y., Liu, X., Wang, K., & He, Q. (2019). A spatio-temporal recommendation model for taxi drivers based on historical GPS data. Transportation Research Part C: Emerging Technologies, 104, 206-221.
22. Feng, Y., Zhang, C., & Zhu, X. (2020). Spatio-temporal recommendation for urban express delivery based on improved collaborative filtering. Computers, Environment and Urban Systems, 81, 101459.
23. Jin, X., Cao, J., & Wang, J. (2019). A spatio-temporal recommendation approach for tour planning. Transportation Research Part C: Emerging Technologies, 102, 105-118.

24. Zhang, M., Chen, X., & Chen, G. (2018). Spatio-temporal recommendation for location-based social networks: a survey. World Wide Web, 21(1), 191-218.
25. Ma, J., Wang, Z., Wu, X., & Yang, X. (2017). Spatio-temporal recommendation in social media. Information Sciences, 381, 70-82.
26. Huang, Z., Lu, X., & Zhang, Z. (2015). Spatio-temporal recommendation in social media. In 2015 IEEE international conference on data mining (ICDM) (pp. 917-922). IEEE.
27. Jannach, D., Zanker, M., Felfernig, A., & Friedrich, G. (2010). Recommender systems: An introduction. Cambridge University Press.
28. Lee, J., & Lee, J. (2018). Spatio-temporal personalized recommendation system for tourist attractions. IEEE Transactions on Industrial Informatics, 14(7), 3296-3304.
29. Liu, H., Yang, X., Zhao, L., & Zhang, X. (2018). A spatio-temporal recommendation algorithm for point of interest based on tourist preferences. In 2018 IEEE international conference on systems, man, and cybernetics (SMC) (pp. 240-245). IEEE.
30. McAuley, J., & Leskovec, J. (2013). Hidden factors and hidden topics: Understanding rating dimensions with review text. In Proceedings of the 7th ACM conference on recommender systems (pp. 165-172). ACM.
31. Nguyen, T. T., & Thai, M. T. (2019). A spatio-temporal recommender system for tourism. Expert Systems with Applications, 131, 104-117.
32. Okoli, C., & Schabram, K. (2010). A guide to conducting a systematic literature review of information systems research. Sprouts: Working Papers on Information Systems, 10(26), 1-52.
33. O'Reilly, T. (2005). What is web 2.0: Design patterns and business models for the next generation of software. Communications & Strategies, 65(1), 17-37.
34. Ribeiro, B., Pereira, F., & Ferreira, A. (2019). A spatio-temporal recommender system for event recommendation. IEEE Access, 7, 90634-90648.
35. Salakhutdinov, R., & Mnih, A. (2008). Bayesian probabilistic matrix factorization using Markov chain Monte Carlo. In Proceedings of the 25th international conference on machine learning (pp. 880-887). ACM.
36. Wang, X., Qin, L., & Zhao, Y. (2019). An intelligent spatio-temporal recommendation approach for tourist destinations. Journal of Ambient Intelligence and Humanized Computing, 10(6), 2199-2210.
37. Zhang, J., Zhou, L., & Liu, Y. (2018). Spatio-temporal personalized recommendation for tourism. In 2018 IEEE international conference on service operations and logistics, and informatics (SOLI) (pp. 18-23). IEEE.
38. Zhou, T., Wang, J., Zhou, Y., & Zheng, K. (2020). A spatio-temporal attention-based method for point-of-interest recommendation. Information Sciences, 511, 290-303.
39. Zhu, X., Zhou, H., & Liu, Y. (2017). A spatio-temporal personalized recommendation algorithm for POI. In 2017 IEEE international conference on web services (ICWS) (pp. 569-576). IEEE.
40. Zou, C., & Sun, Y. (2019). A spatio-temporal personalized recommendation method based on deep learning for location-based social networks. IEEE Transactions on Systems, Man, and Cybernetics: Systems, 50(1).

Chapter 3
Spatial-Based Big Data and Large-Scale Network Management

R. Sheeja, S. Vanaja, Adimoolam M, and Chidambaranathan Bibin

Abstract Handling of location (spatial)- and time (temporal)-based data is enormous since they are represented as images, videos, and audio. There must be challenging to store those data in today's trends, and storage will be hectic in the future since generation and storage are large in volume. In this chapter, it is essential to study large data concerning spatial and temporal data and their networks. Initially, the study initiated the evolution of big and large-scale networks. Later various management techniques were analyzed on big data and large-scale networks concerning spatial.

Furthermore, they essentially studied temporal-based big data and extensive scale network management. Finally, various applications of big data and large-scale networks have been studied, and analytics represented their impact and challenges concerning spatial and temporal data. The phrase "big data" commonly has various ideas, from gathering data from outside sources and storing and preserving it to using analytical methods and tools to analyze the data. It is a popular term in both academics and business. The intent of this section of the book is to give a summary of current big data analytics. Principles to highlight the significance of big data analytics for decision-making applications. To protect data availability, confidentiality, and integrity, it takes various modules to implement security measures for big data. Here are some standard modules used in big data security implementations: In our digital world, security breaches advanced by malicious software (malware) attacks keep growing and pose a significant security risk. Recent malware identification process uses dynamic and static analysis of behavior patterns, and malware

R. Sheeja · S. Vanaja
Department of Information Technology, Easwari Engineering College, Chennai, India
e-mail: r.sheeja@eec.srmrmp.edu.in

A. M (✉)
Department of Computer Engineering, Government Polytechnic College,
Sevvapet, Tamil Nadu, India
e-mail: adimoolam@gptcsevvapet.co.in

C. Bibin
Department of Mechanical Engineering, RMK College of Engineering and Technology,
Chennai, Tamil Nadu, India

J. A et al. (eds.), *Spatiotemporal Data Analytics and Modeling*, Big Data Management, https://doi.org/10.1007/978-981-99-9651-3_3

signatures, take a lot of time, and could be more successful in identifying real-time unknown malware.

The detailed phase can be eliminated using sophisticated machine-learning techniques built in Python. The solution to the current project's issue focuses on machine learning and big data analytics. Malware poses a severe risk to a user's computer system by stealing sensitive data or impeding security. The growth of the smartphone industry has provided malware creators with new opportunities. There is a need to counteract stealth malware strategies since malware varieties are expanding at an astronomical rate every year. This chapter dataset is imported the initial portion of the CSV file in numerical form and sections of the malware image are imported and loaded for better understanding. The machine learning techniques are applied to the dataset in the second portion. The ultimate goal is to provide a novel image-processing method that uses artificial learning techniques and deep learning architectures to achieve a rate of actual accuracy. Our proposed model will outperform conventional machine learning algorithms, according to a comparative analysis of our model. Overall, this work lays the path for the most accurate and efficient malware detection ever.

Keywords Spatial and temporal data · Large-scale network · Large-scale network management · Spatial data large-scale network

3.1 Evolution of Big Data and Large-Scale Networks

Big community records refer to large records that are generated or accumulated in one-of-a-kind networks various from the World Wide Web (WWW), conversation networks, networks, transportation networks, quotation networks, organic networks, etc., are the datasets that have complicated standard information computation techniques are insufficient, while superior facts analytics techniques are wished to extract a fee from them [1–3]. The five most important dimensions of large facts acknowledged as velocity, volume, value, veracity, and variety nevertheless observed huge community data. Big statistics of extensive WI-FI webs have the crucial elements of a large variety, a surplus of volume, real-time velocity, and massive cost leading to special finding complications distinct from the current computing model. In the proposed article, we currently survey the state-of-art huge records analytics tactics for large-scale WI-FI networks [4–6].

This area critiques applicable domain-specific big data challenges in the sequence of how intently they are associated with geospatial concepts and the significance of spatiotemporal wondering in practical answer developments. While big data has the function of 5Vs, the feature-based challenges range in exclusive digital earth applicable domains. The 5Vs characteristics are velocity, variety, veracity, volume, and value. The large scalable networks are social media networks such as Facebook, Twitter, WhatsApp, Telegram, etc. [7, 8]. Since these networks are highly communicating, processing, and storing big volumes of big data, from another angle, large-scale networks are applied for networked surveillance systems such as traffic video

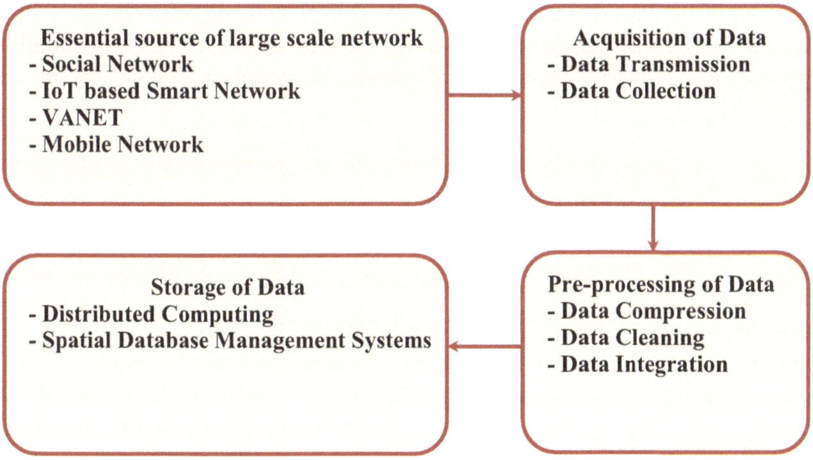

Fig. 3.1 Essentialness of big data and large scale networks

surveillance systems, suspicious people monitoring video surveillance systems, border security surveillance systems, etc. In these surveillance systems, the human being and other objects are watched with event-based surveillance [9, 10].

In recent years, the severe view of the expansion of WIFI conversation technological developments that are widely employed throughout the globe to fulfill the verbal exchange wants from the extraordinarily giant range of cease users and sudden big data [11–13]. The interconnection of several WIFI conversation structures collectively types massive WIFI networks, the place of "large-scale" capacity, the extreme density of community stations, and the giant insurance area. In the interim, there is a sudden increase in the massive extent of cellular records visitors generate from WIFI networks that consist of an enormous variety of wireless gadgets, such as smartphones, cell tags with RFID, tablet computers, and laptops, sensors, clever meters, and clever home equipment [14–19]. Figure 3.1 illustrates large-scale networks' essentialness and ability to generate big data applications. In general, the sources of large-scale networks nowadays are social networks [20–22].

The evolution of huge data sets and networks has been a fascinating and transformative journey, driven by technological advances and the increasing need for processing and analyzing massive amounts of data. Let us explore their evolution in more detail.

3.1.1 Elements of Big Data and Networks

3.1.1.1 Big Data

The concept of a great deal data emerged as organizations started generating and collecting vast amounts of data in the late twentieth century. Initially, data was primarily structured and stored in relational databases. Rise of unstructured data: With the advent of the World Wide Web (WWW), digital technologies and social media data employing text, images, videos, and sensor data began to grow exponentially—traditional data-processing techniques needed to be improved for handling this data deluge [23].

3.1.1.2 Hadoop and Distributed Processing

In the mid-2000s, the Apache Hadoop project introduced a scalable, distributed framework for processing big data. Hadoop's distributed file system (HDFS) and MapReduce processing model enabled parallel processing across clusters of commodity hardware, making it feasible to process large datasets. Evolution of ecosystem: The extensive data ecosystem expanded rapidly with the development of tools and technologies like NoSQL databases, Apache Hive, Apache Spark, and Apache Pig. These frameworks provided more efficient data processing, querying, and storage solutions [24].

3.1.1.3 Large-Scale Networks

The origins of large-scale networks are traced to the early days of computer networking. ARPANET, developed in the late 1960s, laid the foundation for the modern Internet. Internet expansion: In subsequent decades, the Internet expanded globally, connecting millions of computers and devices worldwide. The emergence of protocols like TCP/IP enabled the interconnection of disparate networks into a cohesive network. Web and social networks: The World Wide Web and social media platforms further accelerated the growth of large-scale networks. Websites, online services, and social networking sites facilitated information sharing and collaboration on an unprecedented scale [25].

3.1.1.4 Cloud Computing

The rise of cloud computing systems introduced highly scalable and shared systems. Infrastructure resources that could be deployed and managed on demand were made available by cloud service providers like the Cloud from Google, Azure from Microsoft, and Amazon's own Web Services (AWS). Internet of Things (IoT): The proliferation of devices working in IoT modules added another dimension to large-scale networks. Billions of interconnected devices generate vast amounts of data, enabling real-time monitoring, control, and analytics across various domains [26].

3.1.2 Convergence of Big Data and Large-Scale Networks

3.1.2.1 Information-Driven Insights

The combination of big data and large-scale networks has empowered organizations to gain valuable perceptions from the immense quantity of data produced by networked systems. Machine learning and data analytics techniques are applied to extract knowledge, make informed decisions and detect patterns [27].

3.1.2.2 Personalization and Recommendation Systems

Big data and large-scale networks have fueled the development of personalized services and recommendation systems. By analyzing user behavior and preferences on networked platforms, companies can provide tailored content, products, and recommendations [28].

3.1.2.3 Network Optimization and Security

Analyzing network data can help optimize network performance, identify bottlenecks, and enhance security. Real-time monitoring and analysis of network traffic enable the detection and mitigation of anomalies and cyber threats [29].

3.1.2.4 Social Network Analysis

Big data techniques applied to social network data have led to the emergence of social network analysis. Researchers can study social connections, influence patterns, and information diffusion within large-scale networks, yielding insights into human behavior and societal dynamics. The evolution of big data and large-scale networks continues to unfold, driven by technological advancements, data growth, and the increasing interconnectedness of our digital world. As we move forward, innovations like edge computing, 5G networks, artificial intelligence, and quantum computing are expected to shape the future of big data analytics [30].

3.2 Spatial-Based Big Data and Large-Scale Networks Management

The place viewable data are growing, especially in telecommunication of satellite atmosphere. Almost all developed and developing countries have telecommunication setups and generate large volumes of spatial-based big data with their developed large-scale setup [31]. Currently, with the development of faraway sensors, extensive utilization of GPS gadgets in motors and mobile phones, recognition of geographic data systems, cell applications and crowdsourcing, as properly as affordable computational devices and storage having large information with geographic references is being accumulated from wide fields include ranges from science to engineering, additionally referred to as spatial massive statistics (SBD) [32]. Spatial-based big data [33] and large-scale network management use spatial data and analysis techniques to handle and optimize networks and their associated data. Here is a closer look at these concepts.

3.2.1 Spatial-Based Big Data and Network Management

Spatial data refers to information that represents the physical location and characteristics of objects or phenomena on the Earth's surface or in space. It includes geographic coordinates, maps, satellite imagery, geospatial databases, and 3D models.

3.2.1.1 Integration with Big Data

Spatial data is increasingly being incorporated into the realm of big data. It is combined with other structured and unstructured datasets to enable spatial analysis, visualization, and decision-making on a large scale.

3.2.1.2 Applications of Spatial-Based Big Data

Spatial-based big data finds applications in various domains, including urban planning, transportation management, environmental monitoring, logistics, agriculture, and disaster management. It enables insights into spatial patterns, relationships, and trends that can inform better decision-making processes. Large-scale network management over spatial data: Large-scale networks encompass diverse infrastructure, such as telecommunications networks, computer networks, transportation networks, power grids, and social networks. Managing these networks involves maintaining their stability, efficiency, and security. Spatial aspects: Spatial information plays a crucial role in large-scale network management. It helps identify network components, analyze their geographical distribution, understand connectivity patterns, and optimize resource allocation.

3.2.1.3 Network Performance Optimization

Spatial-based big data can optimize network performance by analyzing spatial relationships between network nodes, identifying bottlenecks, predicting congestion, and optimizing routing and resource allocation based on spatial characteristics.

3.2.1.4 Network Security

Spatial analysis techniques can aid in identifying potential vulnerabilities and threats in large-scale networks. By analyzing spatial patterns of network traffic, anomalies, and suspicious activities, security measures can be strengthened to protect the network infrastructure.

3.2.2 Integration of Spatial Data and Network Management

3.2.2.1 Geographic Information Systems (GIS)

GIS platforms provide tools to manage, analyze, and visualize spatial data. They can be integrated with network management systems to incorporate spatial insights into network operations and decision-making.

3.2.2.2 Real-Time Monitoring

Spatial-based big data allows real-time monitoring of network performance and conditions, leveraging spatial analytics to promptly detect and respond to network events and incidents.

3.2.2.3 Predictive Analytics

By combining spatial data with historical network performance data, machine learning algorithms can predict future network behavior, anticipate network failures, and optimize network operations.

3.2.2.4 Smart City Applications

Spatial-based big data and large-scale network management are the heart of smart city development [34]. They enable the inclusion of various urban systems, such as transportation, energy, public safety, utilities, quality of life, efficiency, and sustainability. Integrating spatial-based big data and large-scale network management opens new opportunities for optimizing network infrastructure, enhancing network performance, and enabling data-driven decision-making in a spatial context.

3.3 Temporal-Based Big Data and Large-Scale Network Management

Temporal-based big data and large-scale network management involve analyzing and controlling time-related data and temporal aspects of networks. Let us delve into the concepts.

3.3.1 Temporal-Based Big Data

3.3.1.1 Time-Stamped Data

Temporal-based big data involves data associated with time stamps, capturing the time of occurrence or measurement. This can include events, transactions, sensor readings, social media posts, and log files that record activities over time.

3.3.1.2 Time-Series Analysis

Time-series analysis techniques analyze temporal patterns, trends, and dependencies within data. It helps uncover insights, detect anomalies, forecast future values, and understand the dynamics of time-varying phenomena.

3.3.1.3 Applications of Temporal-Based Big Data

Temporal-based big data finds applications in diverse fields such as finance, stock market analysis, weather forecasting, IoT sensor data analysis, social media sentiment analysis, healthcare monitoring, and predictive maintenance.

3.3.2 Large-Scale Network Management over Temporal Data

3.3.2.1 Temporal Aspects of Networks

Networks exhibit temporal dynamics, where network states, traffic patterns, and performance parameters change over time. Network management involves understanding and optimizing these temporal aspects.

3.3.2.2 Network Monitoring

Real-time monitoring of network performance metrics, such as latency, packet loss and throughput, helps identify temporal variations, bottlenecks, and congestion in the network. Temporal analysis aids in diagnosing and troubleshooting network issues.

3.3.2.3 Traffic Analysis

Temporal-based big data enables the analysis of network traffic patterns over time. Network capacity planning and resource allocation can be optimized by understanding temporal variations in traffic load, usage patterns, and peak hours.

3.3.2.4 Anomaly Detection

Temporal analysis techniques identify anomalous behavior in network traffic, such as DDoS attacks, network intrusions, or performance degradation. Deviations from standard temporal patterns can trigger alerts and facilitate timely responses.

3.3.2.5 Performance Optimization

Temporal-based big data assists in optimizing network performance by analyzing historical network data. This includes identifying trends, predicting future demands, and adjusting network configurations and protocols accordingly.

3.3.3 Integration of Temporal Data and Management of Network

3.3.3.1 Time-Series Databases

Large volumes of time-stamped data can be stored and queried effectively using time-series databases. They enable high-speed ingestion, indexing, and temporal data analysis for network management.

3.3.3.2 Streaming Analytics

Real-time processing frameworks, such as Apache Kafka and Apache Flink, enable streaming data analysis from network sources. This facilitates real-time decision-making, anomaly detection, and event-driven responses in large-scale networks.

3.3.3.3 Predictive Analytics

Temporal-based big data can be used to develop predictive models for network performance, fault prediction, and capacity planning. Machine learning algorithms can forecast future network behavior and optimize operations by analyzing historical temporal patterns.

3.3.3.4 Network Visualization

Temporal data visualization techniques help represent and understand temporal changes in network behavior. Interactive visualizations and dashboards allow network administrators to manage real-time data. Integrating temporal-based big data and large-scale network management provides valuable insights into network dynamics, performance optimization, and proactive decision-making. Network administrators can better understand network behavior, predict future trends, and ensure efficient and reliable network operations by analyzing temporal patterns.

3.4 Applications of Big Data and Large-Scale Networks

The fast growth of technology has altered daily operations in enterprises and personal lives in this digital environment of Industry 4.0. The evolution of the current notion of the information society has been aided by the Internet of Things (IoT). However, cybercriminals attack personal computers (PCs) and steal personal data to create services for systems and financial gain in industries. Such attackers produce significant system risks and vulnerabilities by using malicious software or malware. Malware is computer software designed to damage the Operating System (OS). Several names, including adware, spyware, and infection, are known as malware. Based on its goal and behavior, it is classified as a command and control (C&C) bot, Ransomware, Rootkit, Worm, Trojan, and Backdoor. Malware detection and mitigation is a fast-developing concern in the cyber security network. Malware authors enhance their ability to elude detection as researchers create new approaches.

Machine learning technology enables the computer to learn automatically from prior data. Machine learning technique uses the mathematical model and creates forecasting through expertise or historical data. It is used for e-mail filtering, captured image recognition, recommender systems, Facebook auto-tagging, speech recognition, and various other tasks. A subset of artificial intelligence (AI) called machine learning enables a machine to learn from past mistakes and on its own from data. The machine learning technique learns from the prior data, builds prediction models, and predicts the result when fresh data is received. The projected output accuracy is improved based on the quantity of amount of data. A vast amount of data assists in constructing a more accurate model that predicts the outcome more precisely. One of the big data-based, large-scale networks is the malware detection system.

The Android operating systems have extremely robust functions and AI-powered applications [2]. Users rely on Android smartphones for various tasks, including making calls, sending texts, playing games, navigating, and doing financial transactions. For malware developers, the growth of the mobile phone community has created modern opportunities. There is a need for stealth malware strategies since malware varieties are expanding at an astronomical rate every year. This study suggests a system for categorizing Android malware processed through machine learning and visualization technique. Images in grayscale were created using Android malware programs from the DREBIN collection. In the initial stage, the suggested framework in Android malware is classified into 15 distinct image portions, later using handcrafted traits found in Android malware images to identify malware files.

The handmade features are extracted from the image sections using methods like local binary pattern (LBP), international image descriptors (ITD), and gray-level co-occurrence matrix-(GLCM). K-closest neighbors and randomly generated forests are two examples of machine learning techniques. The support vector machines further categorized the retrieved features. In the later stage of the experiments, convolutional neural network (CNN) features, and custom features were combined to create a feature fusion technique.

Every segment of a malware image file served as a test case for the categorization performance. The outcomes produced by the feature fusion methodologies are contrasted with those of handcrafted features. The experiment's findings indicate that the feature fusion (FF)-based support vector machine (SVM) model is best among all the Android malware utilizing images of the malware's certificate and Android manifest (CR + AM). A high accuracy of 93.24% was reached. A billion programmed binary files dataset that emerged in 100 million personal computers within a year was introduced in work, and it was found that in a single system, around 94% of these files exist [21]. Since there are many singleton files, malware polymorphism is one reason for this, but there are other causes, as evidenced by the 80:1 ratio of the benign stage to harmful singleton files. It is difficult to detect harmful singletons accurately due to the large number of benign ones. We thoroughly investigate the attributes, traits, and distribution of good and bad singleton files. Most singleton files followed the packing and obfuscation technique for studying the classification, in which 92% of the malicious singletons files have a 1.4% percentage false positive rate.

In code obfuscation techniques, offspring copies are created by altering the parent code having similar serviceability; dissimilar signatures are used to create zero-day or unknown malware [23]. The precision and efficiency needed to identify zero-day malware the approaches outlined in the literature do not now provide. In this research, we have presented and assessed a novel approach to identify zero-day malware that uses several data mining approaches based on the Windows API call frequency. In addition to analyzing the achievement of several data mining algorithms, investigations and analysis were conducted and describes the methodology used to gather the large data sets for training the classifiers. Analyze and consider the benefits of one data mining technique over the other for the successful detection of various zero-day malware by the performance outcomes of these algorithms from our experimental research. The data mining method used in this study picks up knowledge by observing how harmful and benign algorithms behave in sizable datasets.

Here used, several reliable classifiers and assessed for their performance, including k-nearest neighbor (kNN), Naive Bayes (NB), and sequential minimal optimization (SMO) along with kernels (SMO—radiological basis function (RBF); SMO—NormalizedPolyKernel; SMO—PolyKernel; SMO—Puk; J48 decision tree; and the backpropagation technique for neural networks). Overall, the automated data mining technique used in this work produced results with a lower percentage of false positives (FP) of 0.025 and a high. A tremendously favorable rate (TP) greater than 98.5%, which still needs to be seen in the literature. This indicates a significant advancement in identifying zero-day malware because it is significantly higher than acceptable. This chapter also suggests future study directions for examining various obfuscations that now influence the IT industry.

Malware gives risk to computer systems and networks in the way of security that has grown in scope and influence since the infancy of ICT. The volume and diversity of malware variants that are currently present are clear, and traditional prevention measures are mostly unable to handle them [24]. This study looks at how malware has changed over time, including the types of activities it engages in and its varieties, and how that has affected practices in the computer security sector. This chapter proposes a methodology to represent the behavior of its code to extract elements dynamically and statically from malware, like Application Programming Interface (API) calls, as a first step in addressing this difficulty. Machine learning techniques and similarity-based mining have been categorized as malware behaviors. This strategy is based on API sequence calls and frequency of occurrence.

The suggested method effectively distinguishes diverse malware subtypes and categorizes the malware with minimal false alarm rates and high accuracy, according to experimental study results utilizing big datasets. This positive outcome suggests that grouping is a workable method for identifying the similarity to assist in malware identification. The proposed technique guides the researchers to improve the zero-day malware detection process and gives another way to analyze the impact. One of the biggest risks to Internet security right now is malware. It was decided to present a method in this chapter which utilizes MCSC (Malware Classification using SimHash and CNN), a SimHash-based malware classification framework, to turn the disassembled malware codes into grey images. This technique uses a convolutional neural network for classification in their families [25].

Employing techniques like bilinear interpolation, major block, and multi-hash selection enhances the performance of this process. Experimental findings demonstrate that, even for those samples with unequal distribution, MCSC is quite successful in classifying malware families. On a malware dataset of 10,805 samples, the classification accuracy is around 99.260% at its best and 98.862% on average, which is better than other related methods. Additionally, MCSC can recognize a fresh sample in just 1.41 s on average, which is quick enough to fulfil the demands of most real-world applications.

3.4.1 System for Large Network

This chapter proposes a machine learning-based technique for categorizing Android malware. DREBIN dataset is imported for working Android malware applications over a large-scale network. Features from the malware family samples are extracted to identify the features of positive and negative viruses. These features categorize the malware samples present in the DREBIN dataset, which are further classified into their respective malware families. Using machine learning methods like random forest models, machine learning with support vector machines, and k-nearest neighbors, the retrieved characteristics were further categorized. The feature extraction technique is created in the second phase by combining the CNN features with already-in-use algorithms. Every malware image file section was classified based on its performance, and a feature extraction strategy was followed to obtain the result.

3.4.1.1 Access Control Module

This module manages user access to data and resources within the big data environment. It includes authentication mechanisms, role-based access control (RBAC), and access policies to enforce authorized access.

3.4.1.2 Encryption Module

Encryption ensures that sensitive data is protected from unauthorized access. This module employs data encryption at rest and in transit, key management, and encryption algorithms to safeguard the data.

3.4.1.3 Auditing and Monitoring Module

Auditing and monitoring modules track and record misbehaving activities within the big data system. It helps identify suspicious or unauthorized access attempts, monitors data usage, and generates audit logs for compliance and investigation.

3.4.1.4 Data Masking Module

Data masking replaces sensitive data with realistic but fictitious values, making it unreadable to unauthorized users. It preserves data integrity while ensuring confidentiality.

3.4.1.5 Data Loss Prevention (DLP) Module

DLP prevents the unauthorized transmission or disclosure of sensitive data. It uses content filtering, data classification, and policy enforcement techniques to detect and prevent data leakage. Intrusion detection and prevention system (IDPS) module: This module detects and mitigates potential security breaches or attacks. It monitors network traffic, analyzes patterns, and alerts administrators about any suspicious activities in real time. Secure data storage module: This module focuses on securing big data. It includes techniques like secure file systems, secure data backups, data replication, and disaster recovery plans to ensure data availability and resilience.

3.4.1.6 Security Information and Event Management (SIEM) Module

SIEM collects and analyzes security event logs from various sources within the big data ecosystem. It helps identify security incidents, correlates events, and provides centralized monitoring and reporting. Authentication and authorization module: This module manages user authentication and authorization processes. It includes mechanisms like multifactor authentication (MFA), access tokens, and identity management systems to verify user identities and control their access to data.

3.4.1.7 Security Governance and Compliance Module

This module ensures that security measures align with industry standards, regulations, and organizational policies. It involves defining security policies, conducting risk assessments, and implementing compliance frameworks. It is vital to remember that depending on the big data platform, the specific modules and how they are implemented may differ, the organization's requirements, and the regulatory environment.

Figure 3.2 shows the system architecture. The dataset chosen for this project is the DREBIN dataset which contains many malware samples from different families. The malware sample features are then extracted using malware image sections which we refer to as feature extraction. This divides the malware samples into their respective families and helps the dataset remove redundant data. Machine learning algorithms are then trained using the malware dataset with the help of ensemble learning. CNN features are added to the classifier to obtain higher accuracy. The

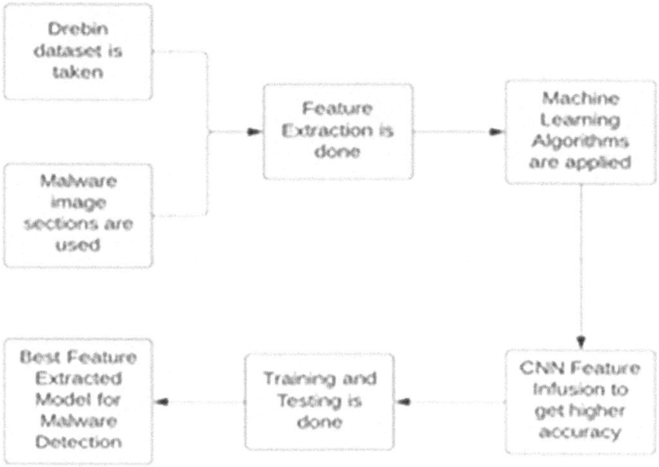

Fig. 3.2 System architecture

final algorithm is then tested and trained for the highest accuracy in detecting Android malware.

3.4.2 *Implementation and Methodology*

The modular design defines the structure of the entire module. Modularity is the degree to which a system's components can be separated into fragments and recombined into a unit.

The proposed system contains the following modules:

1. Data categorization using feature extraction
2. Importing deep learning algorithms
3. Fusion with CNN features
4. Training the algorithm and obtaining the output

3.4.3 *Data Categorization Using Feature Extraction*

Information from a variety of sources is gathered and measured during data classification, illustrated in Fig. 3.3. The acquired data is presented as a CSV file which contains 0 and 1 and is called the DREBIN dataset. The DREBIN dataset consists of two types of malware. The 0s and 1s are converted into images to give higher

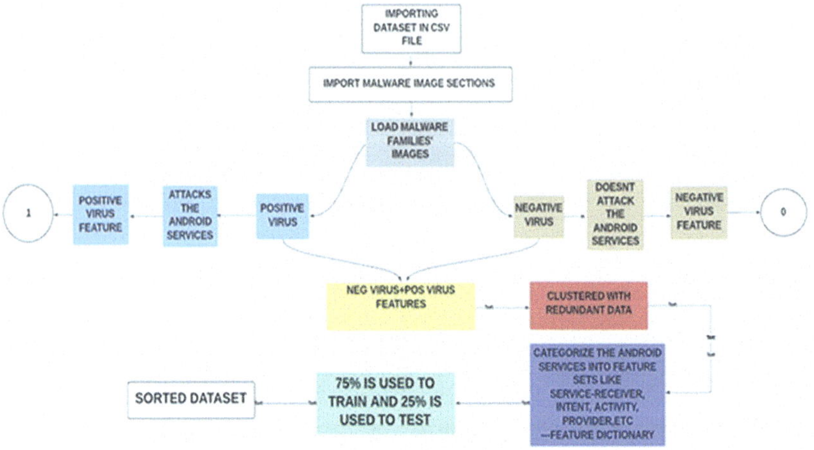

Fig. 3.3 Flow chart of data categorization of using feature extraction

accuracy. If the part of malware attacks the Android services, it is called positive malware (i.e., 1); if it does not affect the Android services, it is called negative malware (i.e., 0). Then the positive malware is considered for training (75%) and testing (25%). At the end of module 1, an organized data set is obtained. If we are interested in exploiting data to build usable AI and machine learning solutions, it needs to be gathered and maintained in an arrangement that makes sense for the specific company's problem [33].

3.4.3.1 Importing Deep Learning Algorithms

The accuracy of the algorithms used in this module, KNN, random forest (RF), and SVM, is tested. Preparing raw data for use with a machine learning model is known as data preprocessing. Making a modelling to facilitate machine learning is the first and most important step. Due to the fact that real-world data frequently contains noise, missing values, and may be in an undesirable format, directly creating machine learning models is not practical. Pretreatment of the data is necessary to clean it and make it suitable for a machine learning model, improving the model's accuracy and effectiveness. The deep learning importation is illustrated in Fig. 3.4.

3.4.4 Fusion with CNN Features

A CNN is a supervised learning technique that may analyze an input image, discriminate between many different items and elements, and assign various entities and features priority (learnable weights and biases). CNNs are used for image

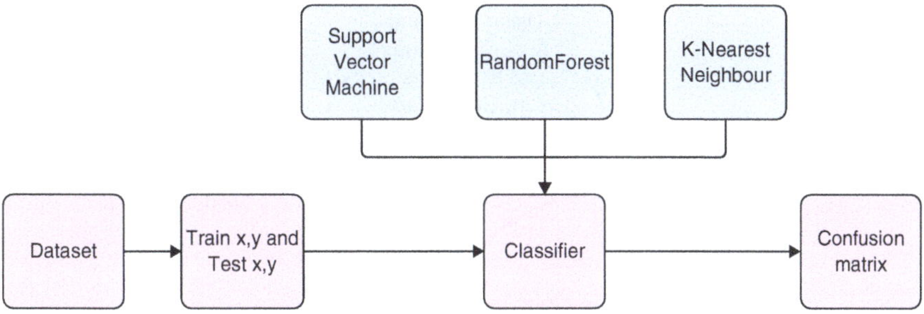

Fig. 3.4 Flowchart of importing deep learning algorithms

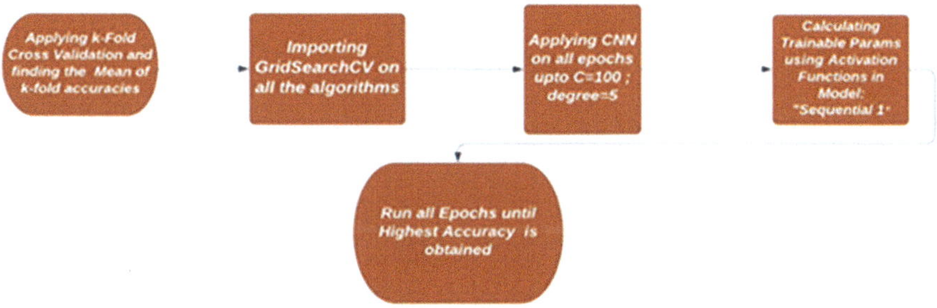

Fig. 3.5 Flowchart of fusion with CNN feature

categorization and recognition due to their high accuracy. The CNN employs a hierarchical network-building method that finally creates a fully connected layer in which all of the neurons are interconnected. The output is processed, as shown in Fig. 3.5.

3.4.5 Training the Algorithm and Obtaining the Output

Training and testing should be configured for the user's comfort while experimenting with fundamental hyperparameters like epoch and learning rate to increase accuracy. Here it is trained to transform the image to grayscale to avoid dealing with a 3D matrix, which is difficult to apply CNN to directly and is not advised for beginners. The set of algorithms that will produce a 99.3% accuracy rating and the output processes of the algorithm depicts this in Fig. 3.6. The experimental result is illustrated in Fig. 3.7.

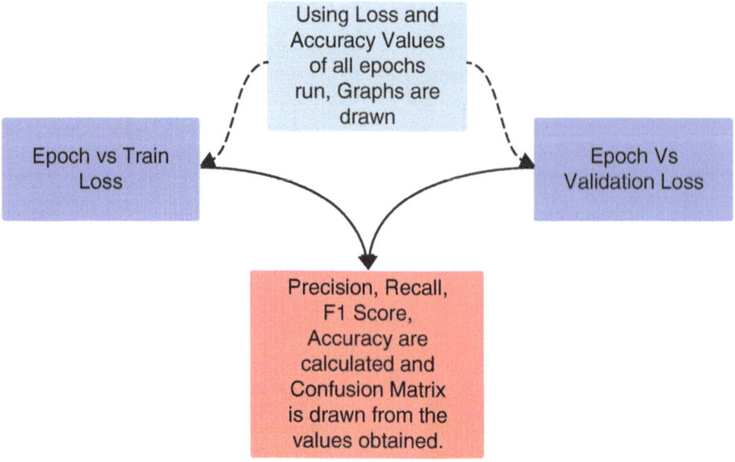

Fig. 3.6 Flowchart of training the algorithm and obtaining the output

```
Epoch 1/5
WARNING:tensorflow:Model was constructed with shape (None, None, 215) for input KerasTensor(type_spec=TensorSpec(shape=(None, None, 21
5), dtype=tf.float32, name='dense_3_input'), name='dense_3_input', description="created by layer 'dense_3_input'"), but it was called o
n an input with incompatible shape (None, 215).
WARNING:tensorflow:Model was constructed with shape (None, None, 215) for input KerasTensor(type_spec=TensorSpec(shape=(None, None, 21
5), dtype=tf.float32, name='dense_3_input'), name='dense_3_input', description="created by layer 'dense_3_input'"), but it was called o
n an input with incompatible shape (None, 215).
356/376 [===========================>..] - ETA: 0s - loss: 0.1107 - accuracy: 0.9632WARNING:tensorflow:Model was constructed with shape
(None, None, 215) for input KerasTensor(type_spec=TensorSpec(shape=(None, None, 215), dtype=tf.float32, name='dense_3_input'), name='de
nse_3_input', description="created by layer 'dense_3_input'"), but it was called on an input with incompatible shape (None, 215).
376/376 [==============================] - 2s 2ms/step - loss: 0.1078 - accuracy: 0.9642 - val_loss: 0.1073 - val_accuracy: 0.9611
Epoch 2/5
376/376 [==============================] - 1s 2ms/step - loss: 0.0573 - accuracy: 0.9828 - val_loss: 0.0404 - val_accuracy: 0.9857
Epoch 3/5
376/376 [==============================] - 1s 2ms/step - loss: 0.0473 - accuracy: 0.9869 - val_loss: 0.0592 - val_accuracy: 0.9810
Epoch 4/5
376/376 [==============================] - 1s 2ms/step - loss: 0.0402 - accuracy: 0.9894 - val_loss: 0.0344 - val_accuracy: 0.9880
Epoch 5/5
376/376 [==============================] - 1s 2ms/step - loss: 0.0350 - accuracy: 0.9910 - val_loss: 0.0299 - val_accuracy: 0.9900
```

Fig. 3.7 Snapshots: training the final algorithm

3.4.6 Test Cases—Accuracy of Classifiers with Traditional ML Algorithms

The initial model is trained with a classifier containing random forest, linear SVC, KNN, and nonlinear SVM algorithms. This initial model was tested against the malware datasets to produce the following accuracy. Accuracy gained as 0.8254073827735284 with the experiment. Rests of the performance measure were also created, as listed in Table 3.1.

Table 3.1 Performance measure of test cases

	Precision	Recall	F1-score	Support	Accuracy
Neg. Virus	0.790	0.988	0.878	1910	0.821
Pos. Virus	0.963	0.542	0.694	1097	0.8214
Macro avg	0.876	0.765	0.786	3007	0.825
Weighted avg	0.853	0.825	0.811	3007	0.8211

3.4.6.1 Applying k-Fold Cross-Validation

A k-fold cross-validation is a data-splitting method when there are $k > 1$ folds. K-fold cross-validation is also known as k-cross, k-fold cross validation, k-fold CV, and k-folds. Python's sci-kit learn to module can be used to implement the k-fold cross-validation technique, which makes calculating k-fold cross-validation models easy. To do model tweaking and choose a model with good generalization performance, it is essential to comprehend cross-validation concepts. When this validation is applied to our existing model and testing it against our malware dataset, we get the following accuracy.

```
[LibSVM]Mean of k-fold accuracies is:  0.8509633431673175
 K-fold accuracies is:  0.03687528952539618
```

3.4.7 Implementing GridSearch CV

GridSearchCV is a technique for choosing the best parameter values from a set of parameters in a grid. It functions basically as a cross-validation approach. It is necessary to provide both the parameters and the model. After obtaining the optimal parameter values, predictions are made. In this model, the proposed model implements the grid search with the parameters as 'C': [0.01, .1, 1, 10, 100]

'kernel' : ['linear', 'rbf', 'sigmoid', 'poly']

'gamma' : [100, 10, 1, 0.1, 0.01, 0.001]

'degree' : [2, 3, 4, 5].

Through the iterations, it is found out that the best accuracy is obtained from 'C': 0.1, 'degree': 2, 'gamma': 1, 'kernel': 'rbf'.

The final accuracy that is obtained after testing it with the malware dataset is:

```
[CV] END ..........C=100, degree=5, gamma=0.001, kernel=poly; total time=   0.2s
[LibSVM]Grid search best accuracy is:  0.9229041916167664
Grid search best parameters are:  {'C': 1, 'degree': 2, 'gamma': 10, 'kernel': 'rbf'}
              precision    recall  f1-score   support

  Neg. Virus      0.952     0.916     0.933      1910
  Pos. Virus      0.862     0.919     0.890      1097

    accuracy                          0.917      3007
   macro avg      0.907     0.917     0.911      3007
weighted avg      0.919     0.917     0.917      3007

Accuracy is:  0.916860658463585
Confusion Matrix is:
[[1749  161]
 [  89 1008]]
```

3.4.7.1 Adding the Keras Optimization

Optimizers use classes or methods that change the weights and learning rate of a deep learning or machine learning model to limit losses. Optimizers help to produce results more quickly. The model is trained for five epochs once the Keras optimizer has been applied, and the final accuracy is as follows. A confusion matrix is also created for this accuracy, as shown in Fig. 3.8

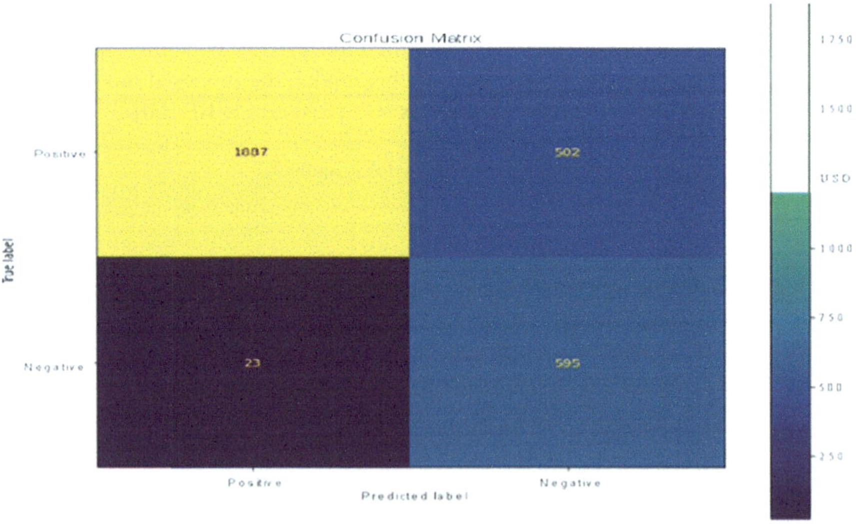

Fig. 3.8 Experimental performance illustration

3.4.8 *Performance Measures of Malware Detection*

3.4.8.1 Precision

Optimizers use classes or methods that change the weights and learning rate of a deep learning or machine learning model to limit losses. Optimizers help to produce results more. In the context of positive observations, precision is defined as the proportion of precisely anticipated observations to all expected positive observations. In other words, accuracy is the classifier's innate ability to avoid classifying a negative sample as positive is depicted in Eq. (3.1).

$$\text{Precision} = \text{TP} / \left(\text{TP} + \text{FP}\right) \tag{3.1}$$

Recall: By reducing the true positives (TP) by the total of true positivity and fake negatives (TP + FN), the recall is calculated. and it is in Eq. (3.2).

$$\text{Recall} = \text{TP} / \left(\text{TP} + \text{FN}\right) \tag{3.2}$$

3.4.8.2 F1 Score

The F1 score is the measured average of Precision and Recall. Therefore, false positives and negatives are considered while calculating this score. It conveys the harmony of recall and precision and is represented in Eq. (3.3).

$$\text{F1 score} = 2^{*}\left(\text{Recall}^{*}\ \text{Precision}\right) / \left(\text{Recall} + \text{Precision}\right) \tag{3.3}$$

3.4.8.3 Support

Support is the proportion of genuine class instances in the provided dataset. Unbalanced support in the training data may lead to the need for stratified sampling or rebalancing and may also be an indication of structural issues with the reported classifier scores. Eq. (3.4) represents the support.

$$\text{Support}\left(\text{X}\right) = \left(\text{Number of transactions containing X}\right) / \left(\text{Total number of transactions}\right) \tag{3.4}$$

3.4.8.4 Accuracy

The total number of samples is divided by the sum of true positives and negatives. If the model is balanced, then this is solely true. If there is a class imbalance, the findings will be erroneous, and it is represented in Eq. (3.5).

$$\text{Accuracy} = \text{TP} + \text{TN} / \left(\text{TP} + \text{TN} + \text{FP} + \text{FN}\right) \tag{3.5}$$

3.4.8.5 Macro Average

Without considering the proportion for each label in the dataset, it computes f1 for each label and then returns the average. The equation for the macro-average F1 score was measured without considering the proportion for each label in the dataset, which can be calculated as follows.

Compute the F1 score for each individual label in the dataset.

$$\text{F1 score for label i} = \left(2^{*}\ \text{Precision\_i}^{*}\ \text{Recall\_i}\right) / \left(\text{Precision\_i} + \text{Recall\_i}\right)$$

Here, Precision_i represents the precision for label i, and Recall_i represents the recall for label i.

Calculate the average of the F1 scores obtained from step 1 for all labels.

$$\begin{aligned}\text{Macro} - \text{average F1 score} = &\left(\text{Sum of F1 scores for all labels}\right) \\ &/ \left(\text{Total number of labels}\right)\end{aligned} \tag{3.6}$$

In this approach, Eq. (3.6), each label is treated equally, regardless of its frequency or proportion in the dataset. When computing the average, the macro-average F1 score gives equal weight to each label, which can be useful in scenarios where you want to evaluate the model's performance across all labels uniformly.

3.4.8.6 Weighted Average

After calculating F1 for each label and taking into account the proportion for each label in the database, the average is returned. The equation for the weighted average F1 score, considering the proportion for each label in the dataset, can be calculated as follows.

Compute the F1 score for each individual label in the dataset.

$$\text{F1 score for label i} = \left(2^{*}\ \text{Precision\_i}^{*}\ \text{Recall\_i}\right) / \left(\text{Precision\_i} + \text{Recall\_i}\right)$$

Here, Precision_i represents the precision for label i, and Recall_i represents the recall for label i.

Calculate the weighted average of the F1 scores obtained from step 1, considering the proportion for each label. The weighted average is represented in Eq. (3.7).

$$\text{Weighted average F1 score} = \left(\text{Sum of}\left(\text{F1 Score\_i}^{*}\ \text{Proportion\_i}\right)\text{for all labels}\right) / \left(\text{Sum of Proportion\_i for all labels}\right) \tag{3.7}$$

Here, F1 Score_i represents the F1 score for label i, and Proportion_i represents the proportion of label i in the dataset. In this approach, each label's F1 score is multiplied by its proportion in the dataset before calculating the average. This ensures that labels with higher proportions contribute more to the final average, reflecting their importance in the overall evaluation.

3.4.9 *Performance Analysis*

The performance of the final trained model is analyzed by plotting the graph between the numbers of epochs and training loss. As the number of epochs increases, the training losses are reduced, and the model's accuracy is increased. It is depicted in the below 2D graphs, as shown in Fig. 3.9.

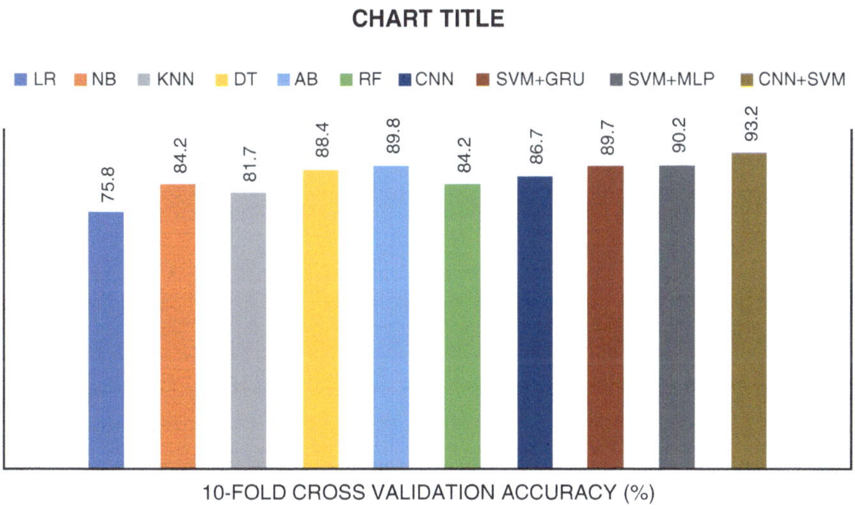

Fig. 3.9 Accuracy of malware detection

3.4.9.1 Epoch vs Validation Loss

Figure 3.10 explains the context of training machine learning models. A complete pass regarding the entire training dataset is often referred to as an epoch throughout the training process. During each epoch, the model processes all the training examples, makes predictions, and adjusts its parameters based on the error or loss incurred.

A complete pass regarding the entire training dataset is often referred to as an epoch throughout the training process. This dataset is distinct from the experimental data and is typically used to assess how well the model generalizes to new, unseen data. During the training process, after each epoch, the model's performance is evaluated on the validation dataset by computing the validation loss. The validation loss measures how well the model performs on the validation data and is usually calculated using a loss function such as mean squared error, cross-entropy loss, or any other appropriate loss function for the specific task. The validation loss serves as a proxy for the model's generalization ability. If the validation loss decreases as training progresses, the model is improving and learning to generalize well. On the other hand, if the validation loss increases or remains stagnant, the model can be poorly generalizing to new data and overfitting the training set. By monitoring the

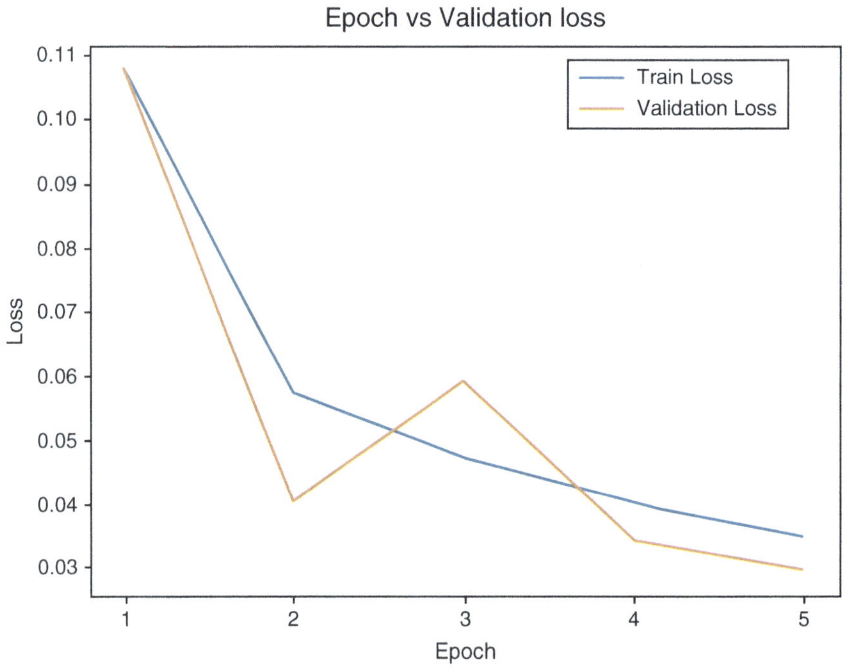

Fig. 3.10 Epoch vs validation loss

validation loss during training, decisions can be made, such as early stopping (to prevent overfitting) or adjusting the learning rate or other hyperparameters to improve the model's performance. The goal is to find a balance where the model achieves low training loss while maintaining a low validation loss, indicating good generalization capabilities.

3.4.9.2 Epoch vs Train Loss

In Fig. 3.11, during the training process, each stage is a whole pass of the trained dataset. During each epoch, the model processes all the training examples, makes predictions, and adjusts its parameters based on the error or loss incurred.

Train loss, or training loss or error, is a statistic used to assess the effectiveness of the model on the training data during each epoch. It quantifies the variation between the values that the model forecasts and the training dataset's actual target values.

The training loss is typically calculated using a function appropriate for the specific task, such as mean squared error, cross-entropy loss, or any other relevant loss function. The loss function quantifies the difference between the predicted and true

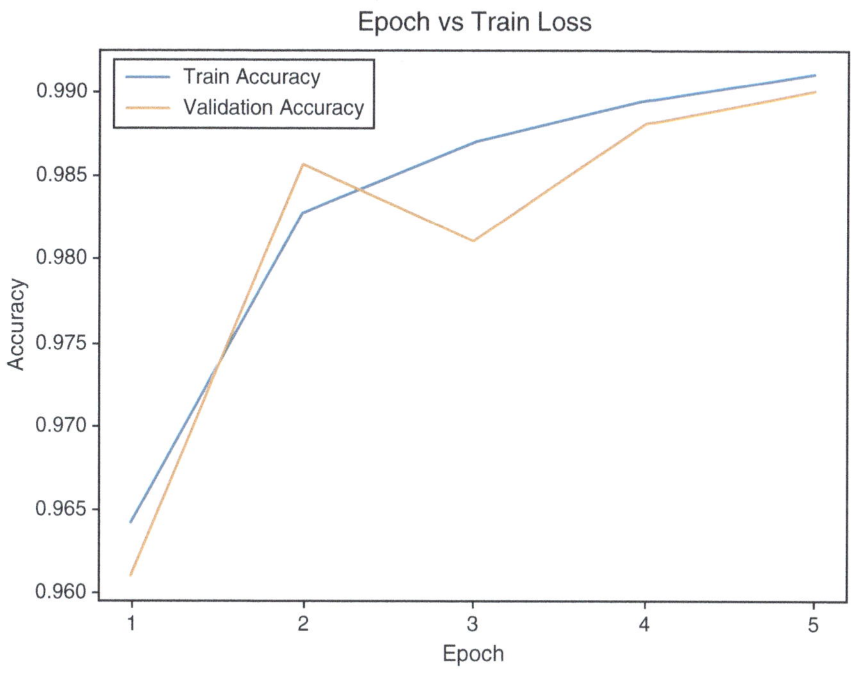

Fig. 3.11 Epoch vs train loss

values, measuring how well the model fits the training data. During the training process, the training loss is monitored to assess the model's performance as it learns and adjusts its parameters. The goal is to minimize the training loss over successive epochs, indicating that the model is becoming more accurate and improving its ability to fit the training data.

It is vital to note that optimizing the train loss alone does not guarantee good generalization to unseen data. Whenever a model overfits the training set of data or becomes excessively complicated, it may achieve a very low train loss but perform poorly on new, unseen data. Therefore, it is crucial to also consider other analysis metrics, such as validation loss or accuracy on a separate validation dataset, to ensure that the model is not overfitting and is able to generalize well to unseen examples. The validation loss helps to assess the model's performance on data that it has not been directly trained on, providing a more reliable estimate of its generalization ability.

3.5 Conclusion

Handling of location (spatial)- and time (temporal)-based data was discussed for handling massive-sized data for large networks since they were represented as images, videos, and audio. The challenge to storing big data was discussed and also, storage hectic was discussed since generation and storage were large in volume. The study of large data concerning spatial and temporal data and their networks was discussed. Initially, the study initiated the evolution of big and large-scale networks. Later various management techniques were analyzed on big data and large-scale networks concerning spatial.

Furthermore, the essentially studied temporal-based big data and large-scale network management were carried out. Finally, various applications of big data and large-scale networks have been studied, and analytics are represented for their impact and challenges concerning spatial and temporal data. This study compares neural networks using deep learning for malware detection for big data-based large network applications with traditional methods for machine learning (MLAs) based on static, dynamic, and image-processing approaches. The system that is being demonstrated conducts malware analysis in a couple of stages and deep learning to examine malware gathered from end-user hosts. In the initial stage, malware categorization was done using a combination of static and dynamic analysis. Using image processing techniques, threat has been broken down into corresponding malware categories in the second stage. In this study, methods that utilized deep learning outperformed traditional MLAs in a number of experimental evaluations utilizing model versions using benchmark datasets that were both publicly and privately available. By integrating a few additional layers to existing designs, the established framework may be scaled up to analyze even more malware in real time and analyze enormous numbers of malware in real time. Future research will concentrate on exploring these variances with elements that are potentially added to the details already accessible.

References

1. Hong-Ning Dai, Raymond Chi-Wing Wong, Hao Wang, Zibin Zheng, and Athanasios V. Vasilakos. 2019. Big Data Analytics for Large-scale Wireless Networks: Challenges and Opportunities. ACM Computer Survey, vol. 52, no. 5, Article 99, pp. 1-36, 2020. doi: https://doi.org/10.1145/3337065
2. Abdulshaheed, H.R., Abbas, H.H., Ahmed, E.Q., Al-Barazanchi, I. (2022). Big Data Analytics for Large Scale Wireless Body Area Networks; Challenges, and Applications. In: Saeed, F., Mohammed, F., Ghaleb, F. (eds) Advances on Intelligent Informatics and Computing. IRICT 2021. Lecture Notes on Data Engineering and Communications Technologies, vol 127. Springer, Cham. doi: https://doi.org/10.1007/978-3-030-98741-1_35
3. Thumati, B.T., Siva Subramania, H., Shastri, R., Kalyana Kumar, K., Hessner, N., Villa, V., et al.: Large-scale data integration for facilities analytics: challenges and opportunities. In: 2020 IEEE International Conference on Big Data (Big Data) [Internet]. IEEE, 2020 December 10. doi: https://doi.org/10.1109/bigdata50022.2020.9378440
4. Sahu, N.K., Patnaik, M., Snigdh, I.: Data Analytics and Its Applications in Brief. Advances in Data Mining and Database Management [Internet]. IGI Global; 2021;115–25. doi: https://doi.org/10.4018/978-1-7998-6673-2.ch008
5. Barazanchi, I.Al., Abdulshaheed, H.R., Safiah, M., Sidek, B.: Innovative technologies of wireless sensor network: the applications of WBAN system and environment. Sustain. Eng. Innov. 1, 98–105 (2020)
6. Ullah, F., Islam, I.U., Abdullah, A.H., Khan, A.: Future of big data and deep learning for wireless body area networks. In: Deep Learning: Convergence to Big Data Analytics. Springer Briefs in Computer Science SCS, pp. 53–77. Springer, Singapore (2019). doi: https://doi.org/10.1007/978-981-13-3459-7_5
7. Shahbazian, R. Distributed optimization over large-scale systems for big data analytics. 4OR-Q J Oper Res 19, 309–310 (2021). doi: https://doi.org/10.1007/s10288-020-00446-x
8. Djedouboum, Asside & ARI, Ado & Gueroui, Abdelhak & Mohamadou, Alidou & Aliouat, Zibouda. (2018). Big Data Collection in Large-Scale Wireless Sensor Networks. Sensors. 18. doi: https://doi.org/10.3390/s18124474.
9. Alsmirat, M.A., Jararweh, Y., Obaidat, I. et al. Internet of surveillance: a cloud supported large-scale wireless surveillance system. J Supercomput 73, 973–992 (2017). doi: https://doi.org/10.1007/s11227-016-1857-x
10. Subudhi, B.N., Rout, D.K. & Ghosh, A. Big data analytics for video surveillance. Multimed Tools Appl 78, 26129–26162 (2019). doi: https://doi.org/10.1007/s11042-019-07793-w
11. L. Li and X. Tan, “Big-Data-Driven Intelligent Wireless Network and Use Cases,” 2021 IEEE International Conference on Communications Workshops (ICC Workshops), Montreal, QC, Canada, 2021, pp. 1-6, doi: https://doi.org/10.1109/ICCWorkshops50388.2021.9473653.
12. C. -L. I, Q. Sun, Z. Liu, S. Zhang and S. Han, "The Big-Data-Driven Intelligent Wireless Network: Architecture, Use Cases, Solutions, and Future Trends," in IEEE Vehicular Technology Magazine, vol. 12, no. 4, pp. 20-29, Dec. 2017, doi: https://doi.org/10.1109/MVT.2017.2752758.
13. Kim, B.-S.; Kim, K.-I.; Shah, B.; Chow, F.; Kim, K.H. Wireless Sensor Networks for Big Data Systems. *Sensors* **2019**, *19*, 1565. doi: https://doi.org/10.3390/s19071565
14. Medeiros, D.S.V., Cunha Neto, H.N., Lopez, M.A. et al. A survey on data analysis on large-Scale wireless networks: online stream processing, trends, and challenges. J Internet Serv Appl 11, 6 (2020). doi: https://doi.org/10.1186/s13174-020-00127-2
15. Capdehourat, G., Bermolen, P., Fiori, M. et al. Large-Scale 802.11 Wireless Networks Data Analysis Based on Graph Clustering. Wireless Pers Commun 120, 1791–1819 (2021). doi: https://doi.org/10.1007/s11277-021-08535-8
16. Challita, U., Ryden, H., & Tullberg, H. (2020). When Machine Learning Meets Wireless Cellular Networks: Deployment, Challenges, and Applications. IEEE Communications Magazine, 58(6), 12–18.

17. Zhang, C., Patras, P., & Haddadi, H. (2019). Deep learning in mobile and wireless networking: A survey. IEEE Communications Surveys Tutorials, 21(3), 2224–2287.
18. Capdehourat, G., Larroca, F., & Morales, G. (2020). A nation-wide Wi-Fi RSSI dataset: Statistical analysis and resulting insights. In: 2020 IFIP Networking Conference, Networking 2020, Paris, France, June 22–26, 2020, IEEE 370–378.
19. Jeunen, O., Bosch, P., Herwegen, M.V., Doorselaer, K.V., Godman, N., & Latré, S. (2018). A machine learning approach for IEEE 802.11 channel allocation. In: 14th International Conference on Network and Service Management (CNSM), pp. 28–36.
20. Asif Iqbal Baba, Hua Lu, Torben Bach Pedersen, and Manfred Jaeger. 2017. Cleansing indoor RFID tracking data. SIGSPATIAL Spec. 9, 1 (July 2017), 11--18.
21. J. Singh, D. Thakur, F. Ali, T. Gera, and K. S. Kwak, "Classification and Analysis of Android Malware Images Using Feature Fusion Technique," IEEEAccess, vol. 09, no. 24, p. 10902, 2021.
22. S.Ni, Q.Qian, R.Zhang,"Malware Identification using visualization images using Machine Learning", IEEE,Vol 77, 871-875, 2018.
23. N. Xie, X. Wang, W. Wang, and J. Liu, Jun. 2019. "Fingerprinting Android malware families," Frontiers Comput. Sci., vol. 13, no. 3, pp. 637–646.
24. M. Tang, M. Alazab, and Y. Luo, "Big data for cybersecurity: Vulnerability disclosure trends and dependencies," IEEE Trans. Big Data, to be published.
25. M. Alazab, S. Venkatraman, P. Watters, and M. Alazab, "Zero-day malware detection based on supervised learning algorithms of API call signatures," in Proc. 9th Australas. Data Mining Conf., vol. 121. Ballarat, Australia: Australian Computer Society, Dec. 2011, pp. 171–182.
26. M. Alazab, S. Venkatraman, P. Watters, M. Alazab, and A. Alazab, "Cybercrime: The case of obfuscated malware," in Global Security, Safety and Sustainability & e-Democracy (Lecture Notes of the Institute for Computer Sciences, Social Informatics and Telecommunications Engineering), vol. 99, C. K. Georgiadis, H. Jahankhani, E. Pimenidis, R. Bashroush, and A. Al-Nemrat, Eds. Berlin, Germany: Springer, 2012.
27. M. Alazab, "Profiling and classifying the behavior of malicious codes," J. Syst. Softw., vol. 100, pp. 91–102, Feb. 2015.
28. S. Huda, J. Abawajy, M. Alazab, M. Abdollalihian, R. Islam, and J. Yearwood, "Hybrids of support vector machine wrapper and filter based framework for malware detection," Future Gener. Comput. Syst., vol. 55, pp. 376–390, Feb. 2016.
29. E. Raff, J. Sylvester, and C. Nicholas, "Learning the PE header, malware detection with minimal domain knowledge," in Proc. 10th ACM Workshop Artif. Intell. Secur. New York, NY, USA: ACM, Nov. 2017, pp. 121–132.
30. C. Rossow, et al., "Prudent practices for designing malware experiments: Status quo and outlook," in Proc. IEEE Symp. Secur. Privacy (SP), Mar. 2012, pp. 65–79.
31. Jiang, Z., Shekhar, S. (2017). Spatial Big Data. In: Spatial Big Data Science. Springer, Cham. doi: https://doi.org/10.1007/978-3-319-60195-3_1
32. R.Y. Ali, V.M.V. Gunturi, Z. Jiang, S. Shekhar, Emerging applications of spatial network big data in transportation, in Big Data and Computational Intelligence in Networking (CRC Press, New York, 2017)
33. Karim Abouelmehdi, Abderrahim Beni-Hessane, and Hayat Khaloufi. 2018. Big healthcare data: Preserving security and privacy. J. Big Data 5, 1 (Jan. 2018), 1.
34. Zhuangbin Chen, Anfeng Liu, Zhetao Li, Young-June Choi, Hiroo Sekiya, and Jie Li. 2017. Energy-efficient broadcasting scheme for smart industrial wireless sensor networks. Mobile Info. Syst. (2017), 1--17.

Chapter 4
Handling Uncertainty in Spatiotemporal Data

Shivaji D. Pawar, Varsha S. Pawar, and Satheesh Abimannan

Abstract Spatial technologies forge massive datasets fast and constantly. This gigantic dataset consists of the time series forecasting or spatial interpolation issue to time and space dimensions. Spatiotemporal data can be further modeled with different statistical, physical, and artificial intelligence (AI) methods, but due to handling uncertainty in spatiotemporal data is the major challenge in front of these models. The chapter's fundamental motivation is to analyze the challenges and strategies for virtually managing uncertainty in spatiotemporal data. The primary difficulties behind the data are high-level feature extractions and long-term memory modeling. These data are technically intensive and result in inadequate model configuration and parameterization. Most AI models oriented with these data need more interpretability and essentially require elaborate training but can model complex nonlinear and Non-Gaussian problems. Predictive uncertainty comes from data and models, which a probability distribution and Bayesian inference could estimate. Therefore, this chapter addresses the detailed strategies for handling uncertainty, including algorithms and approaches for data management. The structure of uncertain data management requires exploring the components of uncertainty management, including data structures and relevant algorithms. This chapter also concentrates on the distinct challenges of handling uncertainty in moving object data and provides strategies for addressing these challenges. Another motivation behind this chapter is to study different domains where spatiotemporal data is encountered on an enormous scale and provides a close look at the computational and I/O requirements of several analysis algorithms for such data. Handling

S. D. Pawar (✉)
Department of Computer Engineering, Anjuman-I-Islam Kalsekar Technical Campus, New Panvel Mumbai, Maharashtra, India

V. S. Pawar
Department of Humanities and Applied Science, Anjuman-I- Islam Kalsekar, Technical Campus, New Panvel, Mumbai, India

S. Abimannan
Amity School of Engineering and Technology, Amity University Maharashtra, Mumbai, Maharashtra, India

J. A et al. (eds.), *Spatiotemporal Data Analytics and Modeling*, Big Data Management, https://doi.org/10.1007/978-981-99-9651-3_4

uncertainty in spatiotemporal data is a hot topic in the research area. This chapter will provide the researchers extensive and revised literature review and future research direction, which will undoubtedly be valuable for addressing the challenges in addressing uncertainty in spatiotemporal data in diverse applications. The view inside the chapter provides state-of-the-art advances in spatiotemporal data handling and highlights new generation necessities to solve uncertainty in spatiotemporal data.

Keywords Spatiotemporal data · Handling uncertainty · Data management AI models

4.1 Introduction to Uncertainty

Spatiotemporal databases deal with data that has both a spatial and temporal component, such as geographical locations with time stamps [1]. These databases are often used in applications like geographical information systems, transportation management, and environmental monitoring. Uncertainty in spatiotemporal databases arises when there is incomplete or inaccurate information about the location or time of a particular event or object [2]. This uncertainty can be caused by several factors, including imprecise measurement tools, incomplete data, or natural variability in the environment being studied [3].

Dealing with uncertainty in spatiotemporal databases is crucial for ensuring the accuracy and reliability of the data being analyzed. One approach to handling uncertainty is to use probabilistic methods, such as Bayesian networks [4], which can incorporate uncertain information into the analysis. Another approach is to use fuzzy logic, which allows for the illustration of imprecise or uncertain information in a more flexible way than traditional binary logic. Fuzzy logic can be used to create rules that can handle uncertain or incomplete data, allowing for more accurate analysis [5]. Overall, dealing with uncertainty in spatiotemporal databases requires careful consideration of the sources of uncertainty and the appropriate methods for handling them. By doing so, researchers and practitioners can ensure that the data they are using is as accurate and reliable as possible, leading to more informed decisions and better outcomes [6].

Spatiotemporal databases are databases that store and manage data that has a spatial and temporal component, such as GPS coordinates with time stamps or weather data over time [7]. One challenge of spatiotemporal databases is dealing with uncertainty in the data. Uncertainty in spatiotemporal data can arise from various sources, such as measurement errors, missing data, data interpolation, and extrapolation. This uncertainty can lead to inaccuracies in data analysis and decision-making. Therefore, it is essential to quantify and manage uncertainty in spatiotemporal databases. One approach to dealing with uncertainty in spatiotemporal data is to use probabilistic models [7]. These models can represent the uncertainty in the data and provide a framework for making probabilistic inferences. For example, a Bayesian network [8] can represent the uncertainty in the data and the relationships between the variables. Another approach is to use fuzzy logic. Fuzzy logic allows for reasoning with imprecise or uncertain data by assigning degrees of truth to statements. For example, a statement like "it is raining heavily" can be assigned a degree of truth between 0 and 1, depending on the level of rainfall.

Finally, data fusion techniques can also be used to deal with uncertainty in spatiotemporal data. Data fusion combines multiple sources of data to produce a more accurate and reliable estimate. For example, combining satellite data with ground-based measurements can provide a more accurate estimate of the temperature or precipitation in an area.

In summary, uncertainty in spatiotemporal databases is a significant challenge, but there are various approaches to deal with it, including probabilistic models, fuzzy logic, and data fusion techniques.

4.2 Spatiotemporal Data Uncertainty

Spatiotemporal data uncertainty refers to the inherent variability and imprecision present in data that involves both spatial and temporal components. It arises due to several factors, including measurement errors, limited data resolution, imprecise location or time stamps, and incomplete or missing data [9]. In spatiotemporal datasets, the spatial component refers to the geographical coordinates or locations, while the temporal component relates to the time stamps or time periods associated with the data points. Uncertainty can arise in either or both of these components, leading to uncertainty in the overall dataset [10]. Figure 4.1 is depicting the factors involves in spatiotemporal data uncertainty.

Dealing with spatiotemporal data uncertainty is a significant challenge in many fields, including environmental monitoring, climate science, transportation, and urban planning [11]. Researchers and practitioners employ various techniques to

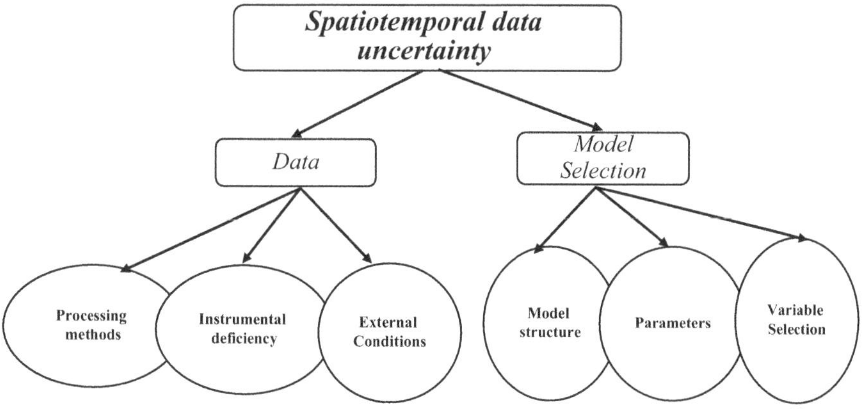

Fig. 4.1 Factors involves in spatiotemporal data uncertainty

account for and manage uncertainty, such as statistical modeling, data fusion, error propagation analysis, and probabilistic approaches [12].

This section highlights the different forms of the data uncertainty:

1. **Spatial uncertainty**: It refers to the lack of pinpoint knowledge about the location or position of a spatial entity or event. This uncertainty can arise from errors in GPS or other positioning systems, limited spatial resolution of data sources, or incomplete knowledge about the boundaries or extent of spatial objects [13].
2. **Temporal uncertainty**: It relates to the lack of precise information about the timing or duration of an event or observation. It can arise from issues such as synchronization errors in data collection devices, delays in data acquisition or transmission, or uncertainty in the occurrence of events [14].
3. **Measurement uncertainty:** It refers to the imprecision or inaccuracy in the measurements themselves. Measurement uncertainty can arise from instrument errors, limitations in measurement techniques, or variations in environmental conditions during data collection [15].
4. **Data quality uncertainty**: It relates to the overall quality and reliability of the data. It can arise from factors such as data incompleteness, missing values, data aggregation or interpolation techniques, or inconsistencies in data sources [16].

Additionally, advanced methods like Monte Carlo simulations, fuzzy logic, or Bayesian inference can be applied to quantify and propagate uncertainty through data analysis and modeling processes, enabling decision-makers to make informed judgments despite the presence of uncertainty in the data.

4.3 Structure of Uncertain Data Management

Uncertain data management refers to the handling and processing of data that is inherently uncertain or imprecise. Data of such type arises in various domains such as sensor networks, scientific simulations, data fusion, and probabilistic databases [16]. To effectively manage uncertain data, several structures and techniques have been developed. Here are some commonly used structures in uncertain data management.

4.3.1 Probabilistic Databases

Probabilistic databases provide a framework for managing uncertain data by incorporating probability distributions over data values. They allow for probabilistic querying and inference, where the results of queries are expressed as probabilities. Here are some key concepts and features of probabilistic databases:

1. **Probabilistic data model**: Probabilistic databases employ a data model that extends the relational or object-oriented data model to handle probabilistic information. They introduce probabilistic tuples or objects that have attributes associated with probability distributions instead of fixed values [17]. These distributions can represent uncertainty, imprecision, or variability in the data.
2. **Probabilistic querying:** Probabilistic databases support querying and inference over the uncertain data. Queries can involve probabilistic conditions, such as "find the most likely objects satisfying a given condition" or "compute the probability of a specific event based on the data." Probabilistic databases utilize probabilistic inference algorithms to evaluate these queries, taking into account the uncertainty in the data [18].
3. **Uncertainty representation:** Probabilistic databases provide mechanisms to represent uncertainty in various forms. They can handle aleatoric uncertainty (inherent variability or randomness in data), epistemic uncertainty (lack of knowledge or imprecision), or both [19]. Common representations include probability distributions, statistical moments, Bayesian networks, or interval-based representations.
4. **Query language extensions:** Probabilistic databases often extend standard query languages, such as SQL, to support probabilistic querying. These extensions include probabilistic operators, such as probabilistic selection, aggregation, and join operations, which take into account the probability distributions associated with data values [20].
5. **Inference algorithms:** Probabilistic databases employ inference algorithms to compute probabilistic query results. These algorithms may involve probabilistic reasoning techniques such as Bayesian inference, Monte Carlo methods (e.g., sampling or Markov Chain Monte Carlo), or exact inference algorithms based on probabilistic graphical models [21].
6. **Probabilistic data integration:** Probabilistic databases can handle data integration scenarios where multiple uncertain or probabilistic data sources need to be combined. Integration techniques involve resolving conflicts, aligning uncertainty representations, and probabilistic fusion of data from different sources [22].

Probabilistic databases find applications in various domains, including sensor networks, uncertain sensor data fusion, scientific data analysis, decision support systems, and machine learning. They enable reasoning with uncertain or incomplete information, handling noisy or imprecise data, and performing probabilistic predictions or inference tasks [23]. It is worth noting that implementing and managing probabilistic databases can be computationally intensive due to the need for probabilistic inference and handling uncertain data. However, they provide powerful tools for working with uncertain or probabilistic information and enable more robust and flexible data analysis.

4.3.2 Interval-Based Representations

Uncertain data can be represented using intervals instead of precise values. Interval-based representations capture the range of possible values for a data item, allowing for reasoning and querying based on intervals. Interval-based representations are a method used to handle uncertainty and imprecision in data by expressing values as intervals rather than precise values [24]. This approach acknowledges that in many practical scenarios, data values may not be known exactly but rather lie within a range or interval. Here are some key points about interval-based representations:

1. **Interval definition**: An interval depicts a range of values between a lower bound and an upper bound. The lower and upper bounds can be inclusive or exclusive, depending on the specific context. For example, an interval [10, 20] represents all values between 10 and 20, including both 10 and 20.
2. **Uncertainty modeling**: Interval-based representations capture the inherent uncertainty or imprecision in data. Instead of assigning a single precise value, intervals provide a way to express a range of possible values [25]. This is particularly useful when dealing with measurements, estimations, or imprecise data sources.
3. **Interval arithmetic**: Arithmetic operations can be performed on intervals to obtain results that incorporate the uncertainty of the operands. Interval arithmetic extends traditional arithmetic operations (such as addition, subtraction, multiplication, and division) to operate on intervals rather than individual values [26]. The resulting interval encompasses all possible outcomes of the arithmetic operation.
4. **Querying and reasoning**: Interval-based representations allow for querying and reasoning based on interval comparisons. For example, querying for objects with attributes falling within a given range, determining overlapping intervals, or assessing the containment of intervals within others. Interval-based reasoning can be applied in various contexts, such as data filtering, data fusion, or decision-making under uncertainty [27].
5. **Interval data types:** Some database management systems and programming languages provide specific data types for interval representation. These data types allow intervals to be stored, manipulated, and queried efficiently. For example, SQL supports interval data types, and programming languages like Python provide libraries for interval arithmetic and handling [28].
6. **Fuzzification:** Interval-based representations can be combined with fuzzy sets to capture both uncertainty and imprecision in data. Fuzzy intervals extend the concept of intervals by associating membership degrees with different values within the interval. This allows for more nuanced modeling of uncertainty and imprecision, considering the degree of confidence or belief in specific values [29].

Interval-based representations find applications in various domains. They are commonly used in scientific measurements, sensor data processing, uncertainty modeling, optimization under uncertainty, and decision support systems. Interval-based techniques also play a crucial role in robust and stochastic optimization, where uncertainty needs to be explicitly accounted for. Interval-based representations provide a flexible and intuitive approach for managing uncertain or imprecise data [30]. They enable reasoning, querying, and computation that considers the range of possible values, allowing for more realistic and reliable analysis and decision-making.

4.3.3 Bayesian Networks

Bayesian networks are graphical models that embody probabilistic relationships among a set of variables. They can be used to model uncertain data and perform probabilistic inference based on the dependencies between variables. Here are some key points about Bayesian networks:

1. **Graphical structure:** Bayesian networks consist of a directed acyclic graph (DAG) where nodes represent variables, and edges represent probabilistic dependencies between variables. The graph structure encodes the conditional dependencies among variables, indicating how one variable influences others [31].
2. **Conditional probability tables (CPTs):** Each node in a Bayesian network is linked with a conditional probability table (CPT) that specifies the probability distribution of the variable given its parents in the graph. The CPT expresses the probabilistic relationship between a variable and its direct causes [31].
3. **Probabilistic inference:** Bayesian networks enable probabilistic inference, allowing for reasoning about the probability distributions of variables given observed evidence. Inference algorithms, such as variable elimination or belief propagation, propagate probabilities through the graph to compute posterior probabilities or make predictions based on observed or queried variables [32].
4. **Learning Bayesian networks:** Bayesian networks can be learned from data using various techniques. Learning algorithms can infer the structure of the network and estimate the parameters in the CPTs based on observed data. Methods such as maximum likelihood estimation, Bayesian learning, or constraint-based approaches can be employed for network learning [33].
5. **Decision-making and utility:** Bayesian networks can be extended to include decision nodes and utility nodes, forming influence diagrams. Decision nodes represent choices or actions, and utility nodes capture the preferences or values associated with different outcomes. This allows for decision-theoretic reasoning, where optimal decisions can be made based on the network's probabilistic relationships and utilities.
6. **Sensitivity Analysis:** Bayesian networks enable sensitivity analysis by assessing how changes in the probability distribution of one variable propagate through the network and affect other variables. This analysis helps comprehend the robustness of the model and identify critical variables or dependencies [34].

Bayesian networks find applications in various domains, including medical diagnosis, risk assessment, financial modeling, machine learning and natural language processing. They are particularly useful in situations with uncertain or incomplete information, allowing for probabilistic reasoning and decision-making. Bayesian networks provide a principled and intuitive framework for representing and reasoning under uncertainty [35]. Their graphical structure facilitates modeling complex probabilistic relationships, and their inference capabilities enable probabilistic reasoning and decision-making.

4.3.4 *Fuzzy Sets*

Fuzzy sets are a mathematical framework that extends the concept of classical (crisp) sets to handle uncertainty and vagueness. Fuzzy sets allow for the representation and alteration of imprecise or ambiguous information, where an element can have a degree of membership in a set rather than a binary membership value of either 0 or 1. Here are some key aspects of fuzzy sets are highlighted in Fig. 4.2 and listed in this section.

1. **Membership function**: Fuzzy sets are defined by membership functions that assign a degree of membership to each element of the universe of discourse. The membership function represents the degree to which an element belongs to the set, ranging from 0 (not a member) to 1 (fully a member). The membership func-

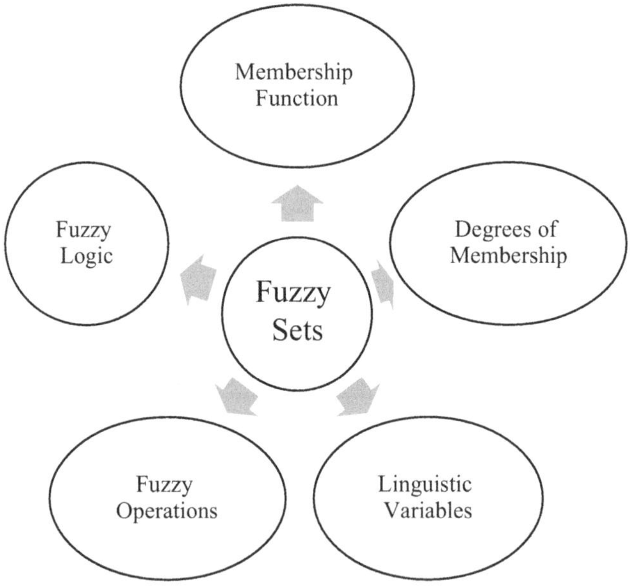

Fig. 4.2 The key components of the fuzzy set

tion can take various forms, such as triangular, trapezoidal, Gaussian, or sigmoidal, depending on the specific context [36].
2. **Degrees of membership**: Unlike classical sets where membership is binary (either an element is in the set or not), fuzzy sets allow for degrees of membership. This means that an element can belong to multiple sets to varying degrees simultaneously. For example, a person's height can be partially tall and partially average, reflecting the fuzzy nature of human perception.
3. **Linguistic variables**: Fuzzy sets are often used to represent and reason with linguistic terms and qualitative information. Linguistic variables express concepts such as "high," "low," "hot," or "cold," which can be imprecise or subjective. Fuzzy sets provide a way to model and manipulate these linguistic terms by associating them with appropriate membership functions.
4. **Fuzzy operations:** Fuzzy sets support operations that generalize classical set operations. Fuzzy set operations include union, intersection, complement, and difference. These operations are defined based on the corresponding operations on the degrees of membership of elements. Fuzzy set operations allow for the combination and comparison of fuzzy sets and their membership degrees [37].
5. **Fuzzy logic:** Fuzzy sets form the foundation of fuzzy logic, a formal system that extends classical Boolean logic to handle uncertain or inaccurate information. Fuzzy logic allows for reasoning and decision-making based on fuzzy sets and their associated operations. Fuzzy logic provides a means to deal with approximate reasoning and approximate solutions.

Fuzzy sets provide a mathematical framework for dealing with uncertainty, imprecision, and vagueness in data and information. By allowing for degrees of membership and linguistic terms, fuzzy sets enable more flexible and nuanced representation, reasoning, and decision-making in the face of incomplete or uncertain information.

4.3.5 Random Sampling

Random sampling is a method used in statistics and research to select a subset of individuals or items from a larger population. The process involves selecting samples in a way that gives each member of the population a like chance of being chosen [38]. Here are some key points about random sampling:

1. **Representative sampling**: Random sampling aims to obtain a representative sample that accurately reflects the characteristics of the entire population. By selecting samples randomly, biases and preferences that could influence the results are minimized, leading to more reliable and generalizable findings.
2. **Simple random sampling**: Simple random sampling is the most rudimentary form of random sampling. In this method, each individual or item in the population has an equal probability of being selected. It can be performed with or without replacement. With replacement means that after an item is selected, it is put

back into the population before the next selection, allowing for the possibility of selecting the same item multiple times. Without replacement means that selected items are not put back, ensuring that each selection is unique.
3. **Random number generators:** Random sampling often relies on random number generators, which generate random numbers within a specified range. These numbers are then used to select the corresponding individuals or items from the population. Random number generators can be computer algorithms or physical devices designed to produce random and unbiased results [39].
4. **Sampling techniques:** Random sampling can be applied using different techniques depending on the characteristics of the population and the research objectives. Some commonly used techniques include simple random sampling, stratified random sampling (where the population is divided into subgroups or strata, and samples are randomly selected from each stratum), cluster random sampling (where the population is divided into clusters, and only a few clusters are randomly selected), and systematic random sampling (where the population is ordered, and samples are selected at regular intervals).
5. **Sample size:** The sample size, or the number of individuals or items selected for the sample, is an important consideration in random sampling. A larger sample size normally leads to more precise and reliable results, but it also requires more resources and time. Statistical techniques can help determine the appropriate sample size based on factors such as the desired level of confidence, variability in the population, and the margin of error allowed.
6. **Statistical inference**: Random sampling forms the basis for statistical inference, where findings from the sample are used to make implications or draw conclusions about the entire population. By ensuring that the sample is randomly selected, statistical techniques can be applied to estimate population parameters, such as means, proportions, or correlations, and assess the uncertainty associated with these estimates [40].

Random sampling is widely used in various fields, including survey research, opinion polls, market research, medical studies, and quality control. By providing an unbiased and representative subset of a population, random sampling helps to obtain reliable and meaningful information for analysis and decision-making.

4.3.6 Monte Carlo Methods

Monte Carlo methods are a class of computational algorithms that use random sampling to solve problems, particularly in the fields of mathematics, physics, statistics, and computer science. These methods rely on repeated random sampling to obtain numerical results, making them particularly useful when exact deterministic solutions are difficult or impractical to calculate. Monte Carlo methods use random sampling to approximate the behavior of complex systems or calculate numerical

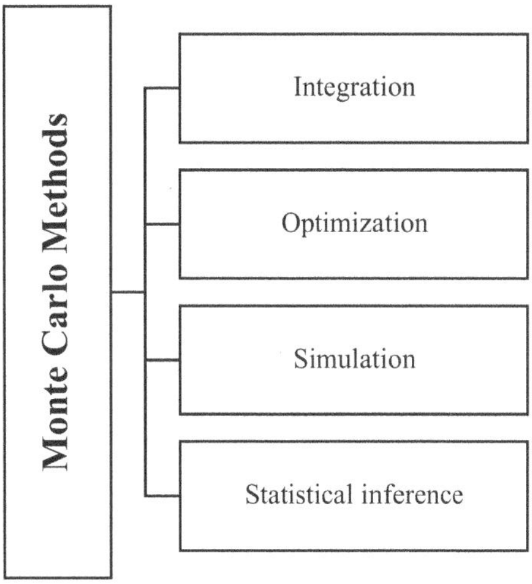

Fig. 4.3 The different Monte Carlo methods for handling data uncertainty

solutions. They involve generating a large number of random samples and using statistical techniques to analyze the results. These methods have a wide range of applications [41, 42] which are depicted in the Fig. 4.3.

1. **Integration**: Monte Carlo integration can be used to numerically evaluate complicated integrals by randomly sampling points in the integration domain. This approach is particularly useful when traditional analytical or numerical integration methods are impractical.
2. **Optimization:** Monte Carlo optimization methods can be utilized to find the global maximum or minimum of a function in a high-dimensional space. By randomly sampling points and evaluating the function at each sample, these methods can converge to an optimal solution.
3. **Simulation:** Monte Carlo simulations are widely used in physics, finance, and other fields to model complex systems and estimate outcomes. By sampling random inputs within specified ranges and using probabilistic models, these simulations can provide understandings into the behavior of the system.
4. **Statistical inference**: Monte Carlo methods are employed in statistics to perform various tasks, such as generating samples from a probability distribution to estimate statistical properties or conducting hypothesis testing.

The effectiveness of Monte Carlo methods relies on the law of large numbers, which states that as the number of samples increases, the average of the samples converges to the expected value [43]. Therefore, by using a large number of random samples, Monte Carlo methods can provide increasingly accurate approximations or estimates. Overall, Monte Carlo methods offer a powerful and flexible approach for tackling problems that are challenging to solve analytically or deterministically. Their ability to harness random sampling and statistical techniques makes them invaluable in numerous scientific and computational fields.

4.3.7 Uncertainty Propagation Techniques

Uncertainty propagation techniques are used to analyze and quantify the effects of uncertainty or variability in input parameters on the output of a mathematical or computational model. These techniques aim to provide insights into the uncertainty of model predictions and assist in decision-making under uncertainty. Here are three commonly used uncertainty propagation techniques highlighted in Fig. 4.4.

1. **Monte Carlo simulation**: Monte Carlo simulation is a statistical technique that involves generating multiple random samples from input parameter distributions and propagating them through the model to obtain a distribution of output values. Each sample is a combination of values drawn from the input parameter distributions, and the model is evaluated using these sampled values [44]. By repeating this process, a large number of times, the resulting distribution of output values provides an estimate of the uncertainty associated with the model predictions.
2. **Analytical propagation**: Analytical propagation methods are used when the relationships between the input parameters and the output of the model can be described using analytical equations. These methods involve calculating the statistical moments (such as mean and variance) of the output variable based on the statistical moments of the input variables. Analytical techniques include methods like the first-order Taylor series approximation, which linearizes the model around the average values of the input variables to estimate the mean and variance of the output [45].
3. **Numerical propagation**: Numerical propagation methods involve approximating the propagation of uncertainty by discretizing the input parameter space and evaluating the model at specific points or through numerical integration techniques. These methods are useful when the model is computationally expensive or lacks a closed-form analytical solution. Examples of numerical propagation techniques include Latin hypercube sampling, quadrature methods, and polynomial chaos expansion [45].

It is important to note that these techniques have their own assumptions and limitations, and the choice of technique depends on the nature of the model and the available information about the input uncertainties. Additionally, advanced techniques like Bayesian inference and Gaussian process regression can also be used for uncertainty propagation in specific contexts.

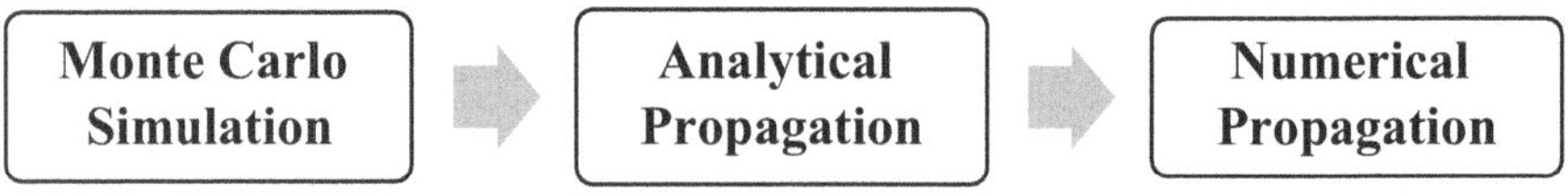

Fig. 4.4 The uncertainty propagation techniques

4.3.8 Stochastic Programming

Stochastic programming is a mathematical optimization framework that addresses decision-making problems under uncertainty. It extends traditional deterministic programming by explicitly considering uncertain parameters in the decision-making process. Stochastic programming models incorporate probabilistic information about uncertain variables to optimize decisions that are robust and take into account the risks associated with uncertainty. In stochastic programming, uncertain parameters are shown as random variables or probability distributions. The objective is to find decision variables that optimize the expected value or some other risk-related measure of the system's performance under uncertainty. There are two main types of stochastic programming models depicted in the Fig. 4.5.

1. **Two-stage stochastic programming**: This model considers a two-stage decision-making process. In the first stage, decisions are made without knowing the realization of indeterminate parameters. In the second stage, after observing the actual values of uncertain parameters, recourse decisions are made to adapt to the realized scenario. The objective is to find a decision strategy that reduces the expected cost or maximizes the expected profit over all possible scenarios [46].
2. **Multi-stage stochastic programming**: Multistage models extend the two-stage model to problems with more than two stages. The decisions are made sequentially over multiple stages, taking into account the evolving information about the uncertain parameters. The objective is to find a decision strategy that optimizes the expected value or a risk measure across all possible scenarios and stages [47].

Stochastic programming has applications in various fields, including finance, energy, transportation, and supply chain management. It provides a powerful framework to make robust decisions in the presence of uncertainty and is particularly useful when the uncertain parameters have a noteworthy impact on the decision outcomes.

These are some of the common structures and techniques employed in uncertain data management. The choice of structure depends on the specific properties of the uncertain data and the requirements of the application at hand.

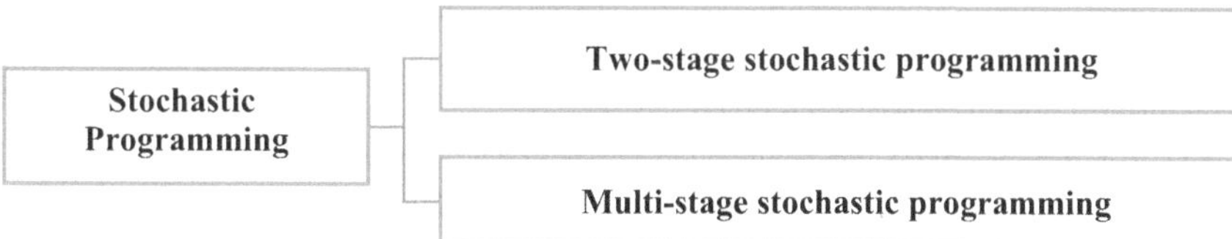

Fig. 4.5 Different types of the stochastic programming

4.4 Moving Object Data Management

Moving object data management involves the storage, retrieval, and processing of data related to objects that are moving in space. This type of data is commonly generated by GPS tracking systems, location-based services, sensor networks, and transportation systems. To effectively handle moving object data, various techniques and approaches have been developed. Here are some key aspects of moving object data management [48].

1. **Trajectory representation:** Trajectories are the paths tracked by moving objects over time. Efficient representation of trajectories is crucial for moving object data management. Common representations include raw coordinate points, polylines, piecewise linear segments, and spline-based representations. The choice of representation depends on the accuracy requirements, storage efficiency, and analysis tasks [49].
2. **Indexing and querying:** Indexing techniques are employed to efficiently store and retrieve moving object data. Spatial indexes, such as R-trees and quadtrees, can be adapted to handle time-varying data by extending the spatial dimensions to include time. Index structures like the spatiotemporal R-tree (STRTree) and the moving object R-tree (MORT) are designed to support efficient spatiotemporal queries, such as range queries, nearest neighbor queries, and trajectory similarity searches [50].
3. **Trajectory compression:** Due to the continuous and high-frequency nature of moving object data, compression techniques are often applied to reduce storage requirements and enable faster data transfer [51]. Compression methods such as Douglas–Peucker algorithm, line simplification, or the use of polynomial fitting techniques can be used to approximate trajectories while minimizing loss of information.
4. **Predictive models:** Predictive models can be employed to forecast future positions or behaviors of moving objects. Techniques such as Kalman filtering, hidden Markov models (HMMs), or machine learning algorithms like recurrent neural networks (RNNs) can be used to predict object trajectories based on historical data. These models enable applications such as real-time tracking, route optimization, and anomaly detection [52].
5. **Spatiotemporal data mining:** Moving object data often requires advanced data mining techniques to notice patterns, trends, and anomalies. Spatiotemporal data mining algorithms, such as clustering, association rule mining, and outlier detection, can be applied to identify significant events, hotspots, or behavior patterns within the moving object data [53].
6. **Privacy and security:** Moving object data often contains sensitive information that needs to be protected. Privacy-preserving techniques, such as anonymization or encryption, can be applied to obfuscate or hide personal information while still allowing useful analysis. Access control mechanisms and authentication protocols are also important for ensuring data security [54].

7. **Stream processing:** Moving object data is typically generated in real-time or streaming fashion. Stream processing techniques, such as stream filters, sliding windows, or stream joins, are employed to efficiently process and analyze data as it arrives in a continuous stream. Stream processing enables real-time monitoring, event detection, and decision-making based on the latest information [55].

These are some of the key considerations in handling moving object data. The specific techniques and approaches used may vary depending on the application domain, data characteristics, and performance requirements.

4.5 Future Scope of Handling Uncertainty

Handling uncertainty in spatiotemporal data is an ongoing and evolving research area with several potential future developments. Here are some potential future directions:

1. Advanced uncertainty modeling: Researchers can explore more advanced uncertainty models that go beyond traditional probabilistic or fuzzy approaches. This includes investigating nonparametric models, Bayesian networks, deep learning-based approaches, or incorporating contextual information to improve uncertainty modeling in spatiotemporal data [56].
2. Uncertainty visualization: Enhancing the visualization techniques for uncertain spatiotemporal data can provide better insights and decision-making support. Future research can focus on developing innovative visual representations, interactive tools, and augmented reality interfaces to effectively communicate uncertainty information to users [57].
3. Real-time uncertainty management: Real-time spatiotemporal data streams, such as social media feeds, IoT sensor data, or mobile device tracking, pose unique challenges for uncertainty management. Future research can focus on developing efficient algorithms and techniques for handling uncertainty in real-time data streams and enabling timely decision-making [58].
4. Uncertainty in machine learning models: Integrating uncertainty handling directly into machine learning models for spatiotemporal data can lead to more robust and reliable predictions. Exploring techniques like Bayesian neural networks, deep ensemble models, or active learning approaches can help improve the performance and interpretability of spatiotemporal machine learning models under uncertainty [59].
5. Integration with domain-specific applications: Applying uncertainty handling techniques in specific domains such as transportation, urban planning, environmental monitoring, or healthcare can lead to domain-specific advancements. Future research can focus on tailoring uncertainty management approaches to the unique requirements and challenges of these applications [60].

4.6 Summary

Handling uncertainty in spatiotemporal data is of paramount importance for accurate analysis, decision-making, and understanding the complexities of dynamic spatial phenomena. The future of uncertainty handling in spatiotemporal data lies in developing more advanced modeling techniques, efficient algorithms, and user-friendly visualization tools. By addressing challenges such as data quality assessment, uncertainty propagation, integration with machine learning models, and real-time processing, researchers and practitioners can enhance the reliability, interpretability, and usability of spatiotemporal data in various domains. Ultimately, advancing the field of uncertainty management in spatiotemporal data will contribute to more informed decision-making, improved risk assessment, and better understanding of dynamic spatial processes, leading to more effective planning, resource allocation, and problem-solving in numerous real-world applications.

References

1. K. L. R. &. R. A. (. N. Toyama, "Geographic location tags on digital images," in *Eleventh ACM international conference on Multimedia*, Redmond WA,, 2003.
2. J. S. Y. M. H. J. &. Z. X. Tang, "Estimating the most likely space–time paths, dwell times and path uncertainties from vehicle trajectory data: A time geographic method," *Transportation Research Part C: Emerging Technologies*, vol. 66, pp. 176-194, 2016.
3. A. e. a. Züfle, "Handling uncertainty in geo-spatial data.," in *IEEE 33rd International Conference on Data Engineering (ICDE)*, San Diego, CA, USA, 2017.
4. Langseth, H., & Portinale, L. (2007). Bayesian networks in reliability. *Reliability Engineering & System Safety*, *92*(1), 92-108.
5. Kosko, B., & Isaka, S. (1993). Fuzzy logic. *Scientific American*, *269*(1), 76-81.
6. Pelekis, N., Theodoulidis, B., Kopanakis, I., & Theodoridis, Y. (2004). Literature review of spatio-temporal database models. The Knowledge Engineering Review, 19(3), 235-274.
7. Chen, L. J., Hsu, W., Cheng, M., & Lee, H. C. (2016, June). LASS: A location-aware sensing system for participatory PM2. 5 monitoring. In Proceedings of the 14th Annual International Conference on Mobile Systems, Applications, and Services Companion (pp. 98-98).
8. Fuhr, N. (1992). Probabilistic models in information retrieval. The computer journal, 35(3), 243-255.
9. Landmann, M., Kaske, M., & Thoma, R. S. (2011). Impact of incomplete and inaccurate data models on high resolution parameter estimation in multidimensional channel sounding. IEEE Transactions on Antennas and Propagation, 60(2), 557-573.
10. Aggarwal, C. C., & Yu, P. S. (2008, April). Outlier detection with uncertain data. In Proceedings of the 2008 SIAM International Conference on Data Mining (pp. 483-493). Society for Industrial and Applied Mathematics.
11. Asariotis, R., & Benamara, H. (Eds.). (2012). Maritime transport and the climate change challenge. Routledge.
12. Morozov, A., & Janschek, K. (2014). Probabilistic error propagation model for mechatronic systems. Mechatronics, 24(8), 1189-1202.
13. Heuvelink, G. B., & Burrough, P. A. (2002). Developments in statistical approaches to spatial uncertainty and its propagation. International Journal of Geographical Information Science, 16(2), 111-113.

14. Rolke, B., & Hofmann, P. (2007). Temporal uncertainty degrades perceptual processing. Psychonomic bulletin & review, 14(3), 522-526.
15. Meyer, V. R. (2007). Measurement uncertainty. Journal of Chromatography A, 1158(1-2), 15-24.
16. Guo, M., & Murphy, R. J. (2012). LCA data quality: sensitivity and uncertainty analysis. Science of the total environment, 435, 230-243.
17. Barbará, D., Garcia-Molina, H., & Porter, D. (1992). The management of probabilistic data. IEEE Transactions on knowledge and data engineering, 4(5), 487-502.
18. Sen, P., & Deshpande, A. (2006, April). Representing and querying correlated tuples in probabilistic databases. In 2007 IEEE 23rd international conference on data engineering (pp. 596-605). IEEE.
19. Costa, P. C. G., Carvalho, R. N., Laskey, K. B., & Park, C. Y. (2011, July). Evaluating uncertainty representation and reasoning in HLF systems. In 14th International Conference on Information Fusion (pp. 1-8). IEEE.
20. Abiteboul, S., Quass, D., McHugh, J., Widom, J., & Wiener, J. L. (1997). The Lorel query language for semistructured data. International journal on digital libraries, 1, 68-88.
21. Cusumano-Towner, M., & Mansinghka, V. K. (2017). AIDE: An algorithm for measuring the accuracy of probabilistic inference algorithms. Advances in Neural Information Processing Systems, 30.
22. Dong, X. L., Halevy, A., & Yu, C. (2009). Data integration with uncertainty. The VLDB Journal, 18, 469-500.
23. Dalvi, N., & Suciu, D. (2007, June). Management of probabilistic data: foundations and challenges. In Proceedings of the twenty-sixth ACM SIGMOD-SIGACT-SIGART symposium on Principles of database systems (pp. 1-12).
24. Allen, J. F. (1981, August). An Interval-Based Representation of Temporal Knowledge. In IJCAI (Vol. 81, pp. 221-226).
25. Ayyub, B. M., & Klir, G. J. (2006). Uncertainty modeling and analysis in engineering and the sciences. CRC Press.
26. Markov, S. M. (1992). Extended interval arithmetic involving infinite intervals. Math. Balkanika, New Ser, 6, 269-304.
27. Tu, S. W., Tennakoon, L., O'Connor, M., Shankar, R., & Das, A. (2008). Using an integrated ontology and information model for querying and reasoning about phenotypes: the case of autism. In AMIA Annual Symposium Proceedings (Vol. 2008, p. 727). American Medical Informatics Association.
28. Walster, G. W. (2005). The use and implementation of interval data types. In Accuracy and Reliability in Scientific Computing (pp. 173-194). Society for Industrial and Applied Mathematics.
29. Rondeau, L., Ruelas, R., Levrat, L., & Lamotte, M. (1997). A defuzzification method respecting the fuzzification. Fuzzy sets and systems, 86(3), 311-320.
30. Selvachandran, G., & Singh, P. K. (2018). Interval-valued complex fuzzy soft set and its application. International Journal for Uncertainty Quantification, 8(2).
31. Friedman, N., & Koller, D. (2003). Being Bayesian about network structure. A Bayesian approach to structure discovery in Bayesian networks. Machine learning, 50, 95-125.
32. Ding, J., & Rebai, A. (2010). Probabilistic inferences in Bayesian networks. Bayesian Network, 39-53.
33. Einhorn, H. J., & Hogarth, R. M. (1985). Ambiguity and uncertainty in probabilistic inference. Psychological review, 92(4), 433.
34. Freeling, A. N. (1981). Alternate theories of belief and the implications for incoherence, reconciliation, and sensitivity analysis. DECISION SCIENCE CONSORTIUM INC FALLS CHURCH VA.
35. Hanea, A., Napoles, O. M., & Ababei, D. (2015). Non-parametric Bayesian networks: improving theory and reviewing applications. Reliability Engineering & System Safety, 144, 265-284.

36. Singpurwalla, N. D., & Booker, J. M. (2004). Membership functions and probability measures of fuzzy sets. Journal of the American Statistical Association, 99(467), 867-877.
37. Zimmermann, H. J. (2010). Fuzzy set theory. Wiley interdisciplinary reviews: computational statistics, 2(3), 317-332.
38. Olken, F., & Rotem, D. (1995). Random sampling from databases: a survey. Statistics and Computing, 5, 25-42.
39. Stipčević, M., & Koç, Ç. K. (2014). True random number generators. Open Problems in Mathematics and Computational Science, 275-315.
40. Cooke, P. (1979). Statistical inference for bounds of random variables. Biometrika, 66(2), 367-374.
41. Hammersley, J. (2013). Monte carlo methods. Springer Science & Business Media.
42. Ferson, S. (1996). What Monte Carlo methods cannot do. Human and Ecological Risk Assessment: An International Journal, 2(4), 990-1007.
43. Kroese, D. P., Brereton, T., Taimre, T., & Botev, Z. I. (2014). Why the Monte Carlo method is so important today. Wiley Interdisciplinary Reviews: Computational Statistics, 6(6), 386-392.
44. Rochman, D., van der Marck, S. C., Koning, A. J., Sjöstrand, H., & Zwermann, W. (2014). Uncertainty propagation with fast Monte Carlo techniques. Nuclear Data Sheets, 118, 367-369.
45. Lee, S. H., & Chen, W. (2009). A comparative study of uncertainty propagation methods for black-box-type problems. Structural and multidisciplinary optimization, 37, 239-253.
46. Barbarosoğlu, G., & Arda, Y. (2004). A two-stage stochastic programming framework for transportation planning in disaster response. Journal of the operational research society, 55(1), 43-53.
47. Vayanos, P., Kuhn, D., & Rustem, B. (2011, December). Decision rules for information discovery in multi-stage stochastic programming. In 2011 50th IEEE Conference on Decision and Control and European Control Conference (pp. 7368-7373). IEEE.
48. Wolfson, O., Xu, B., Chamberlain, S., & Jiang, L. (1998, July). Moving objects databases: Issues and solutions. In Proceedings. Tenth International Conference on Scientific and Statistical Database Management (Cat. No. 98TB100243) (pp. 111-122). IEEE.
49. Hu, X., Han, Y., & Geng, Z. (2021). Novel trajectory representation learning method and its application to trajectory-user linking. IEEE Transactions on Instrumentation and Measurement, 70, 1-9.
50. Sakr, S., & Al-Naymat, G. (2010). Graph indexing and querying: a review. International Journal of Web Information Systems, 6(2), 101-120.
51. Kellaris, G., Pelekis, N., & Theodoridis, Y. (2009). Trajectory compression under network constraints. In Advances in Spatial and Temporal Databases: 11th International Symposium, SSTD 2009 Aalborg, Denmark, July 8-10, 2009 Proceedings 11 (pp. 392-398). Springer Berlin Heidelberg.
52. Waljee, A. K., Higgins, P. D., & Singal, A. G. (2014). A primer on predictive models. Clinical and translational gastroenterology, 5(1), e44.
53. Atluri, G., Karpatne, A., & Kumar, V. (2018). Spatio-temporal data mining: A survey of problems and methods. ACM Computing Surveys (CSUR), 51(4), 1-41.
54. Chanal, P. M., & Kakkasageri, M. S. (2020). Security and privacy in IOT: a survey. Wireless Personal Communications, 115, 1667-1693.
55. Koliousis, A., Weidlich, M., Castro Fernandez, R., Wolf, A. L., Costa, P., & Pietzuch, P. (2016, June). Saber: Window-based hybrid stream processing for heterogeneous architectures. In Proceedings of the 2016 International Conference on Management of Data (pp. 555-569).
56. Radaideh, M. I., Borowiec, K., & Kozlowski, T. (2019). Integrated framework for model assessment and advanced uncertainty quantification of nuclear computer codes under bayesian statistics. Reliability Engineering & System Safety, 189, 357-377.

57. Hullman, J., Qiao, X., Correll, M., Kale, A., & Kay, M. (2018). In pursuit of error: A survey of uncertainty visualization evaluation. IEEE transactions on visualization and computer graphics, 25(1), 903-913.
58. Schneider, M., & Kandel, A. (1994). On uncertainty management in fuzzy inference procedures. Information sciences, 79(3-4), 181-190.
59. Hammer, B., & Villmann, T. (2007, April). How to process uncertainty in machine learning? In ESANN (pp. 79-90).
60. Potter, K., Wilson, A., Bremer, P. T., Williams, D., Doutriaux, C., Pascucci, V., & Johhson, C. (2009, July). Visualization of uncertainty and ensemble data: Exploration of climate modeling and weather forecast data with integrated ViSUS-CDAT systems. In Journal of Physics: Conference Series (Vol. 180, No. 1, p. 012089). IOP Publishing.

Chapter 5
Multimodal Spatial-Temporal Prediction and Classification Using Deep Learning

K. Suresh Kumar, K. Abirami, C. Helen Sulochana, T. Ananth Kumar, Sunday A. Ajagbe, and C. Morris

Abstract "Spatial-temporal" (ST) data is time series data from multiple locations. Data is unpredictable, making predictions difficult. For a variety of urban tasks, such as estimating the speed of traffic and the demand for taxis, an accurate prediction that is based on this kind of data is required. Because of this assumption, the methods that are currently being used that are based on deep learning can only produce one possible outcome. As a consequence of this, they are unable to comprehend how the future will comprise a wide variety of components and how these components will interact with one another. Also, the current method operates under the presumption that information that is spatial and that which is temporal are essentially distinct, and as a result, each must be investigated separately. The chapter introduces a novel approach that utilises spatial-temporal convolutional neural networks in conjunction with Bi-LSTM and enhanced generative adversarial networks (E-GAN) to effectively capture non-linear correlations present in the data distribu-

K. S. Kumar
Sri Krishna College of Technology, Coimbatore, India
e-mail: sureshkumar.k@skct.edu.in

T. A. Kumar (✉)
IFET College of Engineering, Villupuram, India
e-mail: tananthkumar@ifet.ac.in

K. Abirami
Easwari Engineering College, Chennai, India
e-mail: abirami.k@eec.srmrmp.edu.in

C. H. Sulochana
St. Xavier's Catholic College of Engineering, Kanyakumari, Tamil Nadu, India
e-mail: helentcr@gmail.com

S. A. Ajagbe
Ladoke Akintola University of Technology, Ogbomoso, Nigeria
e-mail: saajagbe@pgschool.lautech.edu.ng

C. Morris
V. V. College of Engineering, Tirunelveli, India
e-mail: morrisc97@gmail.com

J. A et al. (eds.), *Spatiotemporal Data Analytics and Modeling*, Big Data Management, https://doi.org/10.1007/978-981-99-9651-3_5

tion. This is achieved through the use of inverse mapping from the forecast distribution. The spatio-temporal correlation network is a modelling technique that captures the distribution of pixels in both space and time. This approach enables the random sampling of latent variables to generate multiple future scenarios. This is accomplished by modelling the spatial distribution of pixels. This sampling can be carried out for a very wide variety of different possible outcomes (STCN). It is a stochastic adversarial network that learns to perform variational inference on data and generate data together with other people through implicit distribution modelling. Education is the means by which one can accomplish both of these goals. E-GAN also allows the combination of external factors, which further improves model learning, and it does this without any additional work. E-GAN outperforms the baseline models and significantly improves performance, as shown by extensive testing on two datasets derived from the real world.

Keywords Spatial-temporal convolutional neural networks · Bi-LSTM Enhanced generative adversarial networks · Spatial-temporal (ST) data · Deep learning-based approaches

5.1 Introduction

The current research focus in the field of computer vision and artificial intelligence is on spatio-temporal (ST) predictive learning, which has been made possible by recent technological advancements [1]. The emergence of these developments has led to the emergence of significant novel research problems. There has been a lot of research done over the past few years into the wide variety of possible applications. Crowd and traffic flow forecasting, weather forecasting, air- and water-quality forecasting, mobile phone usage forecasting, and product demand forecasting are all examples. Cellular traffic and demand forecasting fall under this umbrella as well. Predictive learning using ST models in such contexts is difficult because of the inherent complexity of spatial dependencies and temporal dynamics [2]. For reliable ST forecasts, models must account for the spatial-temporal dynamics between the past and the future. High-dimensional spatial transcriptomics (ST) data can make it challenging to get a clear picture of the dynamics at play. This is because the random events that can occur in such data can be very different from one another. Using deep learning techniques to make spatial and temporal predictions has yielded promising results in recent research. Most literature on the subject states that grids generated by entering latitude and longitude coordinates are the standard way in which urban areas are depicted on maps [3]. In contemporary times, a multitude of scholars have employed a fusion of convolutional neural networks (CNNs) and recurrent neural networks (RNNs) to replicate the spatial and temporal organisation on grid maps. In order to account for the spatial and temporal dimensions of the data, we employed the deep ST approach, which utilises convolutional neural networks [4]. On the other hand, the same thing was carried out with abandoned buildings. Multiple residual convolutional neural networks were used in the study to visualise the spatial data at the hourly, daily, and weekly time scales. Long short-term memory (LSTM) is used in the fields of and in a manner analogous to that of CNN's and GCN's, which first learn to represent spatial information. Yet, in complex environments, accurate spatiotemporal prediction is incredibly difficult to achieve due to a number of obstacles [5].

Vibration-based condition monitoring is widely considered the gold standard for spotting issues. The reason for this is that when components are faulty mechanically they respond differently to external stimulation, which causes vibrations to be transmitted as fault signals [6]. The process of identifying the source of noise in rolling bearings frequently makes use of artificial intelligence techniques that have been in use for some time. However, these methods require pre-selection of features and rely on artificial expertise. As the complexity of the mechanical systems used rises, this task becomes more challenging and less effective. Because of how far computers have come, deep learning can now teach themselves new features. In the past, this skill set was inaccessible [6]. Therefore, deep learning techniques have shown a lot of promise and been extensively researched for use in problem-finding techniques. To better understand how deep learning can be used to diagnose problems in machinery, a summary and plan of attack have been proposed. In order to facilitate intelligent fault diagnosis, a neighbourhood neural network was set up with the aid of a normalised sparse autoencoder. An alternative method of diagnosing electric locomotive bearing problems has been proposed. The vast majority of recent studies on data-driven diagnosis techniques assume the availability and even distribution of large training samples. This means that the deep model needs to be trained with a large enough number of samples with labels to cover a wide range of possible errors. However, this assumption does not always hold true in the engineering world. It is common practice in many sectors to use health-imbalanced samples for training purposes. Most of the time, the best way for rotating machinery systems to function is when they are in good shape and the associated cost and level of safety are taken into account [7]. However, when something does go wrong, the fix usually only lasts for a short while. The collected vibration samples have been found to be adequate in proving the baseline condition. However, the amount of erroneous information linked to out-of-the-ordinary samples is much smaller than that linked to sound samples. A lack of available fault samples makes it more challenging to locate faults when there is a difference between normal and abnormal signals. Numerous academics have developed a wide range of strategies to improve the diagnostic precision currently available, in response to the challenge of fault diagnosis in the presence of unbalanced data. With the help of the extreme learning machine, a sequential prediction method for identifying unbalanced faults was developed. This was done to address the problem of insufficient fault data compared to normal data, which had been collected sequentially online. In order to identify unbalanced historical datasets relating to manufacturing equipment, researchers improved upon the synthetic minority oversampling technique (SMOTE) [8]. To address non-linear issues, the proposed method employs oversampling with a support vector machine (SVM) classifier. For the purpose of classifying samples with varying degrees of information, a deep belief network (DBN) with cost sensitivity was developed and tested on the basis of experiments. The results show that when applied to different datasets, the proposed method outperforms what is already known about fault diagnosis for condition monitoring [9]. Batch normalisation was used to enhance convolutional neural networks (CNNs) that were used to solve the problem of diagnosing problems with rolling bearings. This method accounts for the fact that there are many formats of data and different operational settings. Despite this, most data synthesis techniques are limited in their ability to generate signals and their extraction of complex features. Because of this, they fall short of the numerous demands that training data for intelligent fault diagnosis must meet. Data synthesis has the potential to increase fault diagnosis accuracy in accordance with neural network theory [10]. The generative adversarial network (GAN) is a model that can be used to aggregate a wide variety of data points. The model can generate synthetic signals using random noise, and the resulting data follows the same distribution as the measured vibration signals. Synthesising data can close information gaps for underrepresented groups, altering their historical underrepresentation in databases. Adversarial generative networks (GANs) are in constant competition with one another, making it difficult to keep the two networks in a state of healthy equilibrium. It is not uncommon for the gradient to vanish and the mode to collapse during training, both of which can slow down model convergence and degrade data quality [11]. This has led to the development of many different variations of GAN that are better in terms of both the quality of data synthesis they produce and the ease with which they can be taught. The authors of this study suggest using the Wasserstein distance technique to deal with the problem of mode collapse during the training phase. The objective function also depends on the quality of the synthesised sample. To solve the problem of imbalanced data, you are asked to propose a diagnostic approach that employs deep learning to generate explicit additional training data.

5.2 Analysis of Spatial-Temporal Correlations

5.2.1 Spatial-Temporal Correlations

Spatial-temporal (ST) data is inherently unpredictable and subject to change over time, this scenario exemplifies how the ST data can change significantly over the course of a short period of time [12]. It is crystal clear that both spatial factors, like the area around the PoI, and temporal factors, like the passage of time, have a major impact on the variations in the ST data, especially the outflow trend [13]. Also, things can happen quickly and unexpectedly in a certain place, like blocked driveways, road construction, and traffic jams. These phenomena are not directly observable, but their effects can be seen in the ST data's evolution. The likelihood of such events occurring is influenced by both their geographical location and the time of day they take place. For instance, there are times and places where the likelihood of encountering a traffic jam is higher. Recent methods for predicting ST make use of a layer's worth of spatial and temporal information to make more accurate forecasts. Each mode is represented by a unique model in these methods. Current models are based on the assumption that spatial and temporal factors do not depend on each other very much; this makes it difficult to model changes in data accurately. The models' demonstrated effectiveness under this assumption demonstrates this.

5.2.2 *Multiple Types of Temporal Correlations*

The presented data illustrates the distribution of trips per hour for both the Taxi and Bike datasets. The analysis of the distribution patterns of requests for taxi and bike rides demonstrates noteworthy fluctuations in demand for these transportation modes on a daily, weekly, and even hourly basis [14]. A significant limitation of contemporary techniques pertains to the lack of sophisticated modelling of temporal changes in spatio-temporal (ST) information that is representative of real-world scenarios. The reason for this outcome can be attributed to the utilisation of a unidimensional data mapping methodology, specifically, the hour/day/week model input. The task of identifying the suitable data to incorporate into the model for accurately representing the complex temporal dynamics inherent in the spatio-temporal prediction problem of the real world is a formidable undertaking.

5.2.3 *Multimodality*

Modelling the presence of multiple modes in prediction can be a challenging task, particularly in scenarios involving high-dimensional data and intricate environmental dynamics. Prior spatio-temporal prediction methodologies have operated under the assumption of a deterministic environment, wherein a singular plausible future is present [15]. The assumption in stochastic dynamics can result in suboptimal performance of prediction models in real-world scenarios. A model that employs the mean squared error (MSE) as a metric for evaluating the dissimilarity between predicted and actual data has a tendency to select an average value from a set of slightly different data points. The aforementioned factor results in diminished precision when forecasting over extended periods. According to the findings, a model that is trained using both regression and adversarial loss exhibits superior performance compared to models that solely rely on means. However, it is worth noting that the former model only generates a unimodal distribution, despite the presence of multimodal training data, as previously noted. Thus, it can be posited that standard regressors or classifiers encounter significant challenges in managing multimodality [16].

5.2.4 External Factors Fusion

Forecasting future events can be a challenging task, as spatial-temporal data can be subject to various alterations, leading to intricate modifications in the raw pixel values. The precision of ST forecasting is impacted by a range of factors, including but not limited to temporal considerations, meteorological conditions, unanticipated events, traffic control, and areas of significance [17]. Representation vectors can be concatenated, added, or multiplied individually. The aforementioned solutions are limited in their ability to capture intricate relationships between spatio-temporal data and external factors due to the absence of an objective function for correlation identification. While they can generate a unified representation, the aforementioned solutions are unable to fully capture the complexity of such relationships. Incorporating extraneous variables into spatio-temporal data to enhance model efficacy poses a challenge. The implementation of a deep spatial-temporal generative model (D-GAN) has the potential to enhance the accuracy of spatio-temporal prediction across a diverse set of time intervals, thereby addressing the aforementioned challenges. The D-GAN framework provides comprehensive spatial-temporal forecasting capabilities for urban environments. The D-GAN model is designed to concurrently train the generation and variational inference components, with the aim of achieving accurate stochastic predictions of future events. The D-GAN model acquires knowledge pertaining to stochastic latent variables through the optimisation of a variational lower bound and adversarial loss. This approach facilitates the acquisition of stochastic posterior distributions for spatio-temporal data. The process of making an image can take into account big differences in how the input image looks and how the output image looks. But it might not be enough to deal with differences in space in a good way. The process of video prediction can take into account changes in space, while the way a person looks stays mostly the same from one image to the next. Also, a lot of what you know about the next image in a series of images comes from the image that came before it. Our study involves making VAE-GANs bigger so that they can predict stochastic ST data in urban settings, such as taxi demand and crowd flow, among other things [17]. The proposed model is made up of two different parts: a deep stochastic spatial part and a temporal generative part. The proposed method is to build an enhanced spatio-temporal correlation network that can model the joint distribution of spatial and temporal data well. This network is made to make random sampling of latent variables easier, which makes it possible to come up with many possible futures. Also, the network makes it possible to combine external factors with spatio-temporal data, with the goal of improving the performance of models by getting a shared representation. Then, this shared representation can be used to find connections between the different factors. Next, an implicit distribution modelling-based modified stochastic adversarial network is made to learn about both the generation and variational inference of spatio-temporal data. As far as we know now, our proposal of a latent variable model is a new way to predict stochastic multi-frame spatio-temporal patterns in real-world urban data. This approach has been applied in numerous urban scenarios, including, but not limited to, taxi demand and crowd movement. The utilisation of 3D ST maps with varying temporal resolutions as inputs is recommended for the model. Enabling the model to apprehend intricate temporal dynamics inherent in spatio-temporal (ST) data. Also, our proposed model, E-GAN, has been tested on a large number of real-world datasets, like taxi and bike-sharing data from New York City (NYC).

5.2.5 Spatial-Temporal Prediction in Urban Computing

The implementation of data-driven urban management encounters a significant challenge in the domain of spatial-temporal predictive modelling. In recent times, there has been a proliferation of effective spatio-temporal prediction models that rely on both statistical techniques and deep learning methodologies [18]. In the field of spatio-temporal prediction, the autoregressive integrated moving average (ARIMA) model and its derivatives have been used a lot. Still, these models are limited in their ability to show relationships between space and time. A theory was made in which the expected demand is a combination of three forecasting models, each of which is given a certain amount of weight. Some researchers want to find out how fast people drive their cars and how many cars there are on the road. The methods we have talked about so far are not good enough to predict the amount of traffic in cities because they only look at one or more thoroughfares. Subsequently, prediction methodologies based on time-series are recommended to effectively capture the spatial interdependencies in conjunction with external contextual data such as weather patterns, holidays, and other related factors. However, the existing models were insufficient in elucidating the intricate, non-linear associations among ST. The application of ST data modelling has yielded favourable outcomes in relation to conventional time-series models as well as deep learning models. The researchers leveraged contextual information from various sources and employed a stacked architecture comprising multiple fully connected layers to forecast traffic demand and the disparity between the supply of taxis and the demand for them. The models in question do not establish explicit linkages between the concepts of space and time. A group of researchers conducted an investigation on the convolutional neural network (CNN) to determine its potential for identifying spatial correlations in the context of short-term (ST) prediction. The application of ST data modelling has yielded notable achievements in utilising conventional time-series models as well as models founded on deep learning [19]. The researchers utilised contextual information from various sources and employed a stack of multiple fully connected layers to forecast traffic demand and the disparity between the supply of taxis and the demand from potential passengers. The models in question do not establish clear and direct correlations between spatial and temporal dimensions. A group of researchers conducted an investigation on the convolutional neural network (CNN) with the aim of determining its potential for identifying spatial correlations in the context of spatio-temporal (ST) prediction [20]. Certain scholars employed recurrent neural networks (RNN) to construct a model that captures the temporal dynamics of the data. Recent studies have employed recurrent neural networks (RNNs) and attention mechanisms to acquire knowledge about spatial correlations and temporal dependencies. The gates of the RNN were augmented with ST factors. The ST-ResNet approach leverages the residual neural network architecture to effectively capture the temporal dynamics of urban mobility patterns, including temporal closeness, trend, and periodicity, with the ultimate goal of accurately forecasting the flow of individuals within a given city. The ST network was proposed as a means to forecast demand, with consideration given to both ST temporal correlation and semantic distinctions. A cascade multiplicative unit was incorporated to facilitate the acquisition of knowledge regarding the interdependence of various frames in the prediction of traffic flow. Meta-learning was employed to develop models for predicting traffic patterns based on various geographical factors. A novel framework utilising GAN's was recently introduced for the purpose of forecasting pedestrian flow in a crowded environment.

5.3 Multimodal Spatial-Temporal Prediction and Classification

Multimodal spatial-temporal prediction and classification involve predicting and classifying events that occur over time and space using multiple data sources or modalities. The conundrum at hand can be tackled by employing a combination of CNNs and GANs. GANs have the capability to produce synthetic data instances that closely resemble authentic data. The system comprises of a pair of networks, namely a generator that produces synthetic samples and a discriminator that discriminates between authentic and counterfeit samples. In the context of spatio-temporal prediction and classification, GANs can be used to generate new data points to supplement the existing data set. This can help overcome the problem of limited data availability and improve the accuracy of the predictions.

CNNs are neural networks that use convolutional layers to learn spatial patterns in images or other multidimensional data. They are commonly used for image and video recognition tasks but can also be applied to spatio-temporal prediction and classification. By learning spatial patterns in the data, CNNs can capture the relationships between different features of the data and make accurate predictions. To use GANs and CNNs together for spatio-temporal prediction and classification, a framework can be developed that uses GANs to generate new data samples and CNNs to learn the spatial patterns in the data and make accurate predictions. The framework can be trained on a large dataset that includes multiple modalities, such as images, videos, and other sensor data. The GAN can generate new data samples based on the existing data set, and the CNN can learn the spatial patterns in the data and make accurate predictions.

5.3.1 *Spatio-Temporal Convolutional Neural Networks with Bi-LSTM*

The utilisation of spatio-temporal convolutional neural networks (ST-CNNs) that incorporate bidirectional long short-term memory (Bi-LSTM) layers is a viable approach for conducting multimodal spatio-temporal prediction and classification.

Spatial-temporal convolutional neural networks (ST-CNNs) are a class of neural networks that leverage convolutional layers to acquire knowledge of spatial patterns and temporal variations in data. Video analysis and other spatio-temporal tasks can be performed using them. The utilisation of bi-directional long short-term memory (Bi-LSTM) facilitates the integration of antecedent and subsequent contextual information, as well as their interdependence, into sequences. The utilisation of ST-CNNs in conjunction with Bi-LSTM has been shown to effectively identify spatial and temporal data patterns, resulting in precise predictive outcomes. To use ST-CNNs with Bi-LSTM for multimodal spatio-temporal prediction and classification, a framework can be developed that takes in multiple modalities, such as images, videos, and other sensor data, and processes them using ST-CNNs with Bi-LSTM. The framework can be trained on a large dataset that includes multiple modalities and labels, and the ST-CNNs with Bi-LSTM can learn the spatial and temporal patterns in the data and make accurate predictions.

5.3.2 *Enhanced Generative Adversarial Networks (E-GAN)*

In order to effectively learn about dynamic spatio-temporal correlation, data stochasticity, and the fusion of multi-source data for multi-frame prediction, a novel deep generative model is introduced in the proposed study. The Fig. 5.1 shows a simplified version of the proposed structure for E-GAN which comprises of two components:

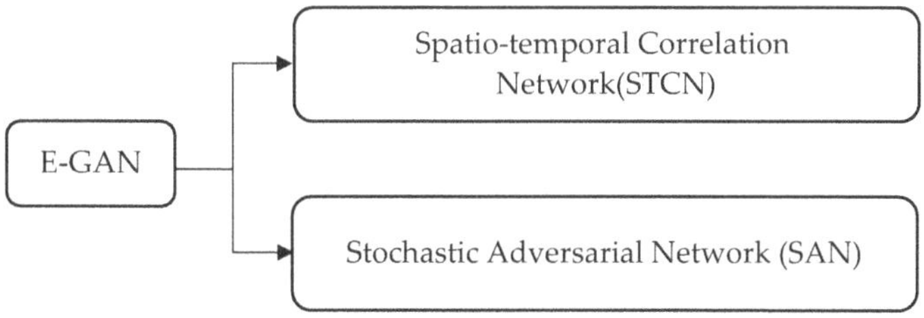

Fig. 5.1 Proposed model components for E-GAN

5.3.3 Enhanced Spatio-Temporal Correlation Network

Enhanced spatio-temporal correlation network (ESTCN) and modified stochastic adversarial network (MSAN) are two different types of neural networks used for various tasks in computer vision and image processing.

The ESTCN model is a neural network architecture specifically developed to capture the spatio-temporal correlations that exist within video sequences. The aforementioned is a recurrent neural network variant that employs a fusion of convolutional and recurrent layers to scrutinise the temporal progression of visual characteristics over time. ESTCN has shown promising results in various applications such as action recognition, video prediction, and video segmentation.

On the other hand, MSAN is a variant of enhanced generative adversarial networks (EGANs) that introduces stochasticity in the generator network to improve the diversity and quality of generated samples. MSAN uses a combination of deterministic and stochastic layers in the generator network, which allows it to generate diverse samples from the same input. MSAN has shown promising results in various applications such as image synthesis, image-to-image translation, and style transfer. Figure 5.2 shows the enhanced spatio-temporal correlation network.

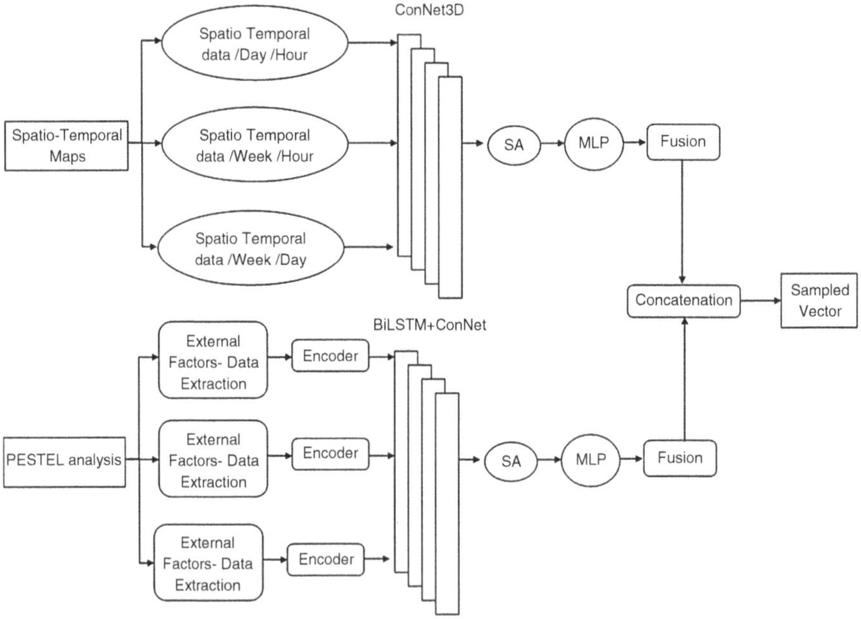

Fig. 5.2 Enhanced spatio-temporal correlation network

5.3.3.1 Encoder

An encoder denoted as E was devised for the purpose of encoding spatio-temporal (ST) data of historical significance, in order to investigate space-time correlation. Analyse the fusion network of the encoder. The BiLSTM and Convolution layer are utilised to extract features from ST data. The BiLSTM architecture incorporates a convolution layer. The BiLSTM model employs a convolution layer to substitute the matrix multiplication process at each gate with a convolution operation. The aforementioned approach is employed in datasets with a high number of dimensions to precisely detect spatial characteristics. The bidirectional long short-term memory (BiLSTM) neural network architecture is designed to incorporate both current input and past states of neighbouring elements in order to effectively capture and represent temporal dependencies. The convolutional neural network (CNN) and bidirectional long short-term memory (BiLSTM) layers are capable of handling spatial and temporal dependencies, respectively.

$$I_t = \rho\left(G_{\text{xi}} \times \chi_t + \text{M}_{\text{hi}} \times \text{Z}_{t-1} + \Gamma_{\text{cf}} * R_{t-2} + G_i\right) \tag{5.1}$$

$$F_t = \rho\left(G_{\text{xf}} \times \chi_t + \text{M}_{\text{hf}} \times \text{Z}_{t-1} + \Gamma_{\text{cf}} * R_{t-2} + G_f\right) \tag{5.2}$$

$$H_t = R_t \tanh\left(\phi_t\right) \tag{5.3}$$

The above equation shows how to use the "*" convolutional operator, where the variables i0, f0, C0, o0, and H0 represent the input gate, forget gate, cell state, output gate, and hidden state at time t, respectively. The input gate's job is to control the information that will be processed, while the forget gate's job is to decide which information should be thrown away. Last, the output gate is in charge of choosing the final output. The symbol for the input gate is i0y], and the symbol for the forget gate is f0y]. The state of the cell is shown by the symbol C0y], and the output gate is shown by the symbol o0y]. Also, H0y stands for the hidden state at the previous time step $t - 1$. The sigmoid and hyperbolic tangent functions are the same thing as the and tanh activation functions. In the context of this study, the variable W stands for "weight," and the variable "b" stands for "bias."

Several stacks use convolutional bidirectional long short-term memory (ConvBiLSTM) units to pull out spatial and temporal features. The output of the convolutional bidirectional long short-term memory (ConvBiLSTM) model goes through a series of changes, including a three-dimensional convolutional neural network (3D-ConvNet), a self-attention layer, and a multi-layer perceptron (MLP), which produces a condensed feature vector called V_x.

$$V_x = \text{MLP}(\text{Attention}(\text{ConvNet3D}(\text{PConvBiLSTM}(\text{ConvBiLSTM}^{\text{L}}\left(X_{\text{real}}\right) \tag{5.4}$$

$$V_y = \text{MLP}(\text{Attention}(\text{ConvNet3D}(\text{QConvBiLSTM}(\text{ConvBiLSTM}^{\text{M}}\left(X^{m}{}_{\text{real}}\right) \tag{5.5}$$

$$V_z = \text{MLP}(\text{Attention}(\text{ConvNet3D}(\text{RConvBiLSTM}(\text{ConvBiLSTM}^{\text{N}}\left(X_{\text{fake}}\right) \tag{5.6}$$

The convolutional bidirectional long short-term memory (ConvBiLSTM) neural network has demonstrated efficacy in detecting long-term patterns in ST map sequences. Conversely, the 3D convolutional neural network (3D-ConvNet) exhibits proficiency in detecting proximal spatial interdependencies. Research has demonstrated that the utilisation of 3D-convolutional neural networks (3D-ConvNets) facilitates the monitoring of transient variations in demand and crowd movement. The observed enhancement can be attributed to the preservation of the proximity of input points through weight sharing across diverse locations in the input, as well as the utilisation of spatio-temporal locality in feature representations. Research has indicated that the utilisation of 3D-convolutional neural networks (3D-ConvNets) enhances the model's capacity to generalise. The utilisation of a self-attention layer facilitates the effective capture of spatial dependencies that are distant in nature. Forecasting spatio-temporal data in urban environments, such as estimating the required number of taxis or the expected crowd size, may necessitate establishing linkages between geographically distinct regions within a metropolis. The utilisation of the convolutional approach is prevalent in contemporary methodologies to demonstrate interdependence among diverse urban regions. The convolution operator possesses a limited receptive field, thereby necessitating multiple convolutional layers to discern long-range dependencies.

5.3.3.2 Fusion

To enhance the model's efficacy, the integration of diverse external factors is employed. The efficacy of spatio-temporal (ST) applications, including but not limited to traffic and demand prediction, is subject to the impact of various external factors, such as points of interest (PoI), weather conditions, and temporal dynamics. A decision has been made to establish a fusion network with a broader scope, aimed at incorporating diverse external factors from multiple domains. The auxiliary feature vector that has been acquired is denoted as V_{ExFi}, where i is a variable that ranges from 1 to w, with w representing the overall number of external factors that are intended to be employed. The process of concatenating the condensed feature vectors has been found to enhance the precision of feature representation, thereby resulting in an overall improvement in system performance.

$$V^{c}_{ExF} = \sum_{i=0}^{\infty}\left(\left(V_{ExF^{\alpha}}\right)\oplus V_{ExF^{\alpha+1}}\right) \tag{5.7}$$

The symbol "$\oplus$ "is used to indicate a sequence of interconnected elements. Encoders which are capable of extracting features from external factors' information are commonly referred to as ExF encoders. The fusion of the spatial-temporal feature representation denoted as V is utilised as input for the purpose of generating the shared representation **V§**. The distribution of ST connection across individual pixels cannot be observed in the entire latent vector Vs. The data is distributed uniformly over every individual pixel.

Modified Stochastic Adversarial Network

MSAN is a variant of enhanced generative adversarial networks (EGANs) that introduces stochasticity in the generator network to improve the diversity and quality of generated samples. The MSAN has shown in Fig. 5.3. MSAN uses a combination of deterministic and stochastic layers in the generator network, which allows it to generate diverse samples from the same input. MSAN has shown promising results in various applications such as image synthesis, image-to-image translation, and style transfer.

The enhanced generative adversarial networks (EGANs) are a variant of the popular generative adversarial networks (GANs) that uses a modified architecture to improve the stability and quality of the generated images.

In EGANs, the generator network uses a modified architecture known as the "skip connections" which allows the network to preserve low-level features of the input image while generating high-level features. This is achieved by introducing shortcut connections between the encoder and decoder layers of the generator network, which helps to transfer information about the low-level features of the input image directly to the decoder layers. Additionally, EGANs use a feature matching loss function, which encourages the generator network to generate images that match the statistics of the real images in terms of their intermediate feature representations. This helps to improve the overall quality and realism of the generated images.

The process of obtaining the integrated and cohesive V_r representation of the ST and ExFs data necessitates the primary utilisation of a multi-layer perceptron (MLP) on the multimodal V representation. The reconstructed spatio-temporal map is derived by employing a multi-layer perceptron (MLP), stacked convolutional long short-term memory (ConvLSTM) layers, and a three-dimensional convolutional neural network (3D-ConvNet) as proposed by G, subsequent to the decoding procedure of the deeply fused representation.

$$X_{enc} = \text{ConvNet3D}\left(\text{PConvBiLSTM}^{R}\left(V_r\right)\right) \tag{5.8}$$

The reconstructed ST map of **xme/î** is denoted as *xean*, while the reconstructed map of **x¬AÆ** is represented as **xean¬AÆ**. In addition, the generator (G) receives a

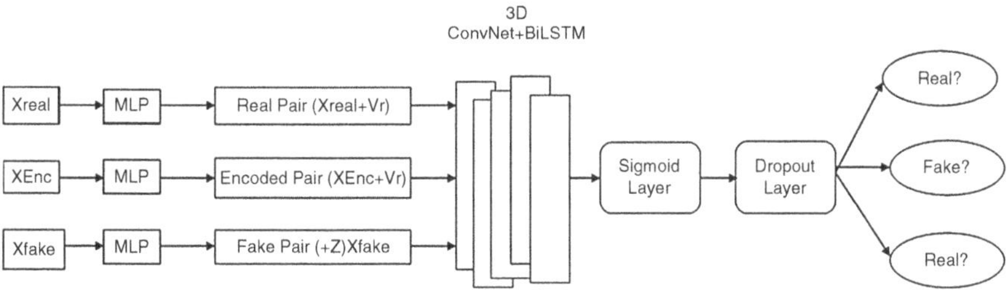

Fig. 5.3 Modified stochastic adversarial network

noise vector (**z**) as an input to produce a reconstructed spectrotemporal (ST) map of the noise vector **x**~/°e. Following the completion of training, the generator function G is utilised to produce synthetic samples that exhibit similarities to authentic data.

Dropout: The individual denoted as 'D' acquires the ability to assess the spatial-temporal (ST) map of any given methodology, encompassing those of 'G' as well as the actual ground truth. The input to D is obtained by concatenating the deep fused representation V_r with the generated ST map latent xean, as well as the real ST map latent xme/i. It is possible to acquire knowledge of both the latent code and data space concurrently. This approach enhances the rate at which convergence occurs, facilitates the acquisition of knowledge, and promotes uniformity in the training process. Specifically, we encode the pairs as *yean* = [**xean**¥,***r***] and *yme/î* = [**xme/î**¥,**V*r***]. A fabricated pair, *y*~/°e = [**x**~/°e¥,], is also produced, wherein **z** represents the feature vector for noise. Subsequently, comparable stacked ConvLSTM layers and a ConvNet3D layer are employed in the following manner:

$$D_{\text{out}} = \sigma\left(\text{ConvNet3D}\left(\text{ConvLSTM}^{\text{R}}\left(y\right)\right)\right) \tag{5.9}$$

The variable *y* represents three distinct quantities, namely yme/î, yean, and y~/°e. The sigmoid function, denoted by σ(•), is utilised to transform the convolutional output into a probability. The variable DÇ¶0 represents the anticipated likelihood of the input variable *y* being classified as either genuine or counterfeit. The training of G and D is conducted concurrently until D reaches a point of inability to differentiate between the ST map produced by G and the ST map generated from authentic data.

5.4 Analysis of Spatial-Temporal Prediction and Classification

Using two comprehensive real-world datasets, this research shows the model's efficacy in forecasting taxi service demand, a difficult task. The objective of the demand prediction task is to construct a precise model capable of forecasting the aggregate quantity of service requests anticipated in every grid of a spatio-temporal (ST) map within a specified time period. There is a time difference of 168 h from the initial 60-min interval. This study expands the framework for crowd behaviour prediction. The findings suggest that the proposed solution provides a flexible framework for forecasting spatio-temporal phenomena in various urban settings. The primary objective of the task of predicting crowd flow is to develop a precise model capable of forecasting the aggregate count of individuals entering and exiting each grid on the spatio-temporal map at specified time intervals. This will enable the accomplishment of the task's additional objectives.

5.4.1 Datasets Description

NYC TLC Trip Record Data: The datasets are taken from the website https://www1.nyc.gov/site/tlc/about/tlc-trip-record-data.page. These platforms provide access to various datasets related to NYC, including taxi trip data. Every travel record comprises of the latitude and longitude coordinates as well as the timestamps of the pickup and drop-off occurrences. This dataset includes detailed information about taxi trips, such as pickup and drop-off locations, time stamps, fares, trip distances, and other relevant attributes. It covers both yellow taxis and green taxis (also known as Boro Taxis) operating in NYC.

5.4.2 CitiBikeNYC Dataset

The CitiBikeNYC dataset contains information about bike-sharing trips in New York City. CitiBike is a bike-sharing programme in NYC operated by Motivate (now a part of Lyft). The dataset provides details about the trips taken by CitiBike users, including information such as trip durations, start and end times, start and end stations, bike IDs, user types, and more.

To access the CitiBikeNYC dataset, you can visit the official CitiBike system data webpage (https://www.citibikenyc.com/system-data) or explore the NYC open data platform (https://opendata.cityofnewyork.us/) as they often host the dataset. Figure 5.4a and b represent the clusters distribution of TaxiNYC and CitiBikeNYC dataset. The dataset is typically available in CSV format and can be downloaded for

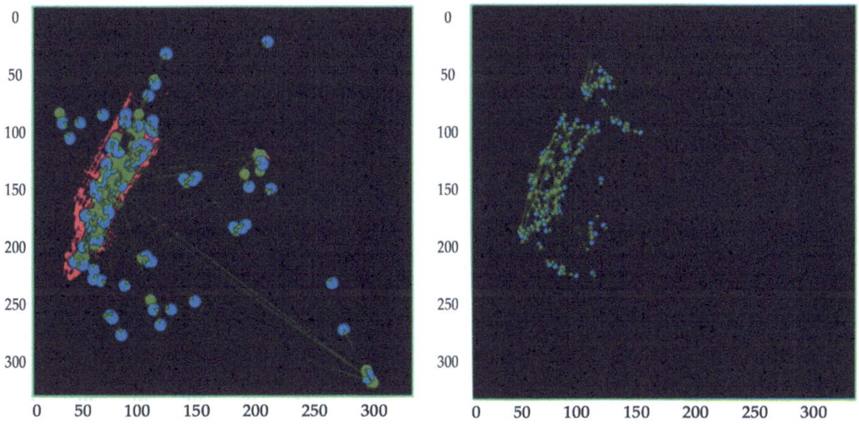

Fig. 5.4 (**a**, **b**) Clusters distribution of TaxiNYC and CitiBikeNYC dataset

analysis and research purposes. Commercial centres, transportation hubs, and business districts are some of the examples of distinct areas that can be found within metropolitan areas. These distinct areas serve a variety of functions. There are a variety of different patterns of traffic flow and taxi demand across the various functional domains. For instance, students leave their homes in the morning to travel to school, and then they travel back to their homes in the evening. It has been discovered that some already existing datasets segment the New York City metropolitan area into regions without taking into account the dynamics of those regions. Because of this, certain locations are referred to as "water areas," and the demand there is always considered to be zero. This can lead to a reduction in the algorithm's mean error, which can have a significant impact on how the algorithm's overall performance is evaluated. In light of this, we investigated the various functional domains included in the NYC dataset. Following this, we applied the Kmeans clustering method to divide the NYC area into 100 distinct areas, each of which was determined by the average historical demand in that particular area. Following that, the region that has consistently low demand is removed from the list, bringing the total number of active regions up to 81.

We have included the (16, 8, 10) map from the widely used pre-processed BikeNYC dataset in our analysis because we want to demonstrate that our proposed solution is feasible on a more granular level. This map is used to predict crowd flow and comes from the dataset that is widely used. Both the input ST map and the output ST map in the given dataset have different dimensions. The input ST map has the values (16, 8) and the output ST map has the values (16, 8, 2). Therefore, the solution that we have proposed has been modified to include the conditional E-GAN model. In this model, the output of G is conditional on what it is fed.

Furthermore, the integration of historical data with external variables, such as weather data, points of interest (PoI), and the differentiation between weekends and weekdays, is undertaken. Furthermore, we utilise publicly available New York3 point of interest (PoI) data. The total number of points of interest (POIs) is 18,912, which includes many facility domains. The training of the model is presently being carried out at the University Research Facility in Big Data Analytics (UBDA), which is situated at Hong Kong Polytechnic University. The process involves the utilisation of two Tesla P100-PCIE GPU. Each graphics processing unit (GPU) has a memory capacity of 16 gigabytes, and the underlying operating system is specifically engineered to accommodate 64-bit architecture. The memory's operational frequency is 1.3285 gigahertz. The official online platform for accessing information related to the University Big Data Analytics Research Centre (UBDA) is available at the web address https://www.polyu.edu.hk/ubda/. The programming environment utilised in this research is Keras, with TensorFlow serving as the foundational framework.

5.4.3 Performance Comparison with Baselines

This section provides an assessment of E-GAN in comparison to traditional methods employed for predicting the demand volume on the NYC taxi and Bike datasets. Table 5.1 displays a comparative analysis of the performance of E-GAN in contrast

Table 5.1 Performance comparison among different baseline methods

	NYC TLC trip record data		CitiBikeNYC dataset	
Models	RMSE	MAE	RMSE	MAE
LSTM	12.31	0	6.54	7.23
CNN	14.21	4.01	10.12	9.12
STRESNET	7.54	3.43	8.34	8.23
DMVST	4.43	4.12	6.23	9.34
STGCN	5.34	2.12	5.43	6.54
D-GAN	1.42	1.12	4.32	2.54
E-GAN	1.21	0.97	3.11	1.02

to other baseline techniques. Table 5.1 illustrates the primary observations that can be deduced. The empirical results indicate that the E-GAN model outperforms alternative approaches in the domain of demand forecasting. The taxi and bike datasets exhibit the lowest root mean squared error (RMSE) values of 1.21 and 3.11, respectively, as well as the lowest mean absolute error (MAE) values of 0.97 and 1.02, respectively. The results demonstrate a notable increase of 49.2% in relative improvement compared to the optimal performance among all baselines for the taxi dataset, and a significant increase of 39.4% compared to the optimal performance among all baselines for the bike dataset. The proposed methodology attains the minimum root mean square error (RMSE) of 4.77 and the minimum mean absolute error (MAE) of 2.63 when implemented for the purpose of forecasting crowd flow. The experimental results demonstrate a notable enhancement of 20.23% (RMSE) relative to the optimal performance observed across all baseline models. It is posited that the advantages can be ascribed to the effective architecture of the E-GAN, which was capable of concurrently capturing the dynamic spatial and temporal characteristics of the data, in addition to the multimodal nature of the information. The conventional techniques, namely, LSTM and CNN, exhibit unsatisfactory outcomes, as evidenced by the RMSE scores of 12.31 and 6.54, correspondingly, for taxi data, and 14.21 and 10.12, respectively, for bike data.

Although regression techniques such as RF and XGB have demonstrated superior performance compared to their predecessors, they are limited in their ability to simultaneously capture the inherent spatio-temporal variation trend present in the data. The incorporation of well-established deep learning methodologies, namely BiLSTM and CNN, broadens the extent of our model comparison. Although both BiLSTM and CNN demonstrate constraints in representing spatial correlations, CNN is notably inadequate in capturing temporal dynamics. The ST-ResNet and DMVST-Net methodologies are widely acknowledged as the foremost benchmark techniques for forecasting grid-centric data. Two distinct methodologies are utilised with the aim of constructing models that account for spatial correlations. The long short-term memory (LSTM) and convolutional neural network (CNN) methodologies exhibit certain benefits from the aforementioned techniques, albeit with specific constraints. On the other hand, it can be observed that STGCN presents only slight improvements when compared to LSTM and CNN. This is due to the fact that STGCN utilises previous iterations' expected conditions as past observations for subsequent iterations. As a result, the process of prediction resulted in the accumulation of inaccuracies. The likelihood of error aggregation escalates with a rise in the quantity of layers within the network architecture, specifically in the domain of extended-term prediction. In addition, traditional machine learning methods produce results using the mean squared error (MSE) metric, which calculates a pixel-level average across multiple slightly different possible solutions within the pixel space. The inadequacies of mean squared error (MSE) in the context of modelling multimodal data arise from its incapacity to furnish a solution that accounts for multiple modes. The methodologies mentioned earlier incorporate spatial and temporal features in a straightforward manner, resulting in their inadequacy to adequately capture the inherent correlations between spatial and temporal characteristics. Furthermore, the consolidation of extraneous variables within a specific site prevents any possible association between the ST data and those external factors. The study proposes the E-GAN model, which possesses the capacity to acquire comprehensive data distributions implicitly. This capability enables the assimilation of inherent connections among spatial and temporal characteristics and the construction of models for multiple modes of data. This is in contrast to alternative models. Table 5.1 illustrates that E-GAN exhibits superior precision when compared to traditional and deep learning-oriented ST prediction models. The aforementioned statement underscores the benefits of acquiring a comprehension of spatial and temporal correlations, integrating external factors via an objective function, and proficiently handling data multimodality. The E-GAN model exhibits superior precision in ST prediction when compared to conventional and deep learning-oriented models for ST prediction. In contrast, E-GAN exhibits the ability to prevent the propagation of inaccuracies at the pixel level due to its inherent acquisition of the data distribution. This study pertains to the forecasting of 24 frames in every subsequent iteration. The results indicate that the performance of the suggested methodology surpasses that of the conventional approaches in evaluating 24 forthcoming frames. The inadequacy of current methodologies in capturing the time-varying patterns of spatio-temporal content can be attributed to their reliance on two-dimensional spatio-temporal maps as their primary input. The approach we propose differs in that it involves utilising 3D ST maps with varying temporal resolutions as input, which allows for the modelling of complex temporal dynamics. The utilisation of 3D spatio-temporal maps, commonly referred to as video, enables the capture of complex spatial relationships that undergo temporal variations.

5.4.4 Analysis of Variants of E-GAN

The project entails a comprehensive examination of the fundamental components of our multimodal E-GAN to gain a deeper understanding of its operational mechanisms. Two inquiries were posed by individuals: To what extent does the design of the encoder and decoder impact the efficacy of capturing dynamic spatial and temporal dependencies? Does the selection of the loss function impact the accuracy of predictions? To address the aforementioned inquiries, five distinct iterations of the proposed methodology have been considered.

The present investigation employs adversarial learning techniques and long short-term memory (LSTM) layers to proficiently model spatio-temporal data. There exist a total of 64 concealed units.

The research employs an adversarial learning approach utilising the GANs-CNN architecture, comprising of two convolutional layers and one fully connected layer.

- The method under consideration employs adversarial learning, utilising a combination of two CNN layers, a fully connected layer, and a LSTM layer.
- The GANs-ConvLSTM-n method is founded on the fundamental structure of GANs. It comprises of four convolutional long short-term memory (ConvLSTM) layers and one three-dimensional convolutional neural network (3DConvNet) layer, all of which have identical parameters. The architectural structure of the edifice bears resemblance to that of E-GAN.

The research employs a fundamental structure GANs that comprises four convolutional long short-term memory (ConvLSTM) layers and one three-dimensional convolutional neural network (3DConvNet) layer, which possess identical parameters. The present architectural configuration exhibits similarities to D-GAN in its ability to effectively manage the stochastic nature of the data. The dissimilarities can be categorised as the subsequent components: The present investigation solely employs STp data as input data, which is denoted as a three-dimensional tensor. Each spatio-temporal (ST) map depicted in the tensor represents the ST data pertaining to a specific hour within a singular day. It is noteworthy that the analysis did not incorporate any extraneous variables. The L2 loss in the D-GAN-FS model is replaced by the feature similarity loss.

Table 5.2 shows how the performance of E-GAN compares to the performance of other variants. It can be seen that the complete version of the E-GAN model shows better performance in most of the evaluation metrics in a variety of situations. In particular, we give a summary of three important points:

The results depicted in Table 5.2 demonstrate that generative adversarial networks (GANs) exhibit superior performance compared to existing deep learning

Table 5.2 Comparison between different variants of E-GAN

	NYC TLC trip record data		CitiBikeNYC dataset	
Models	RMSE	MAE	RMSE	MAE
GANs-LSTM	6.81	4.988	5.56	6.34
GANs-CNN	12.28	4.408	9.14	8.23
GANs-CNN-LSTM	9.61	5.098	7.36	7.34
GANs-ConvLSTM-n	8.61	3.098	5.25	8.45
GANs-ConvLSTM-s	5.01	8.208	4.45	5.65
D-GAN-FS	6.283	7.261	3.34	1.65
D-GAN	5.556	2.432	3.65	2.14
E-GAN	1.21	0.97	3.11	1.02

models, owing to their capacity to acquire knowledge pertaining to sample generation from probability distribution. The GANs-LSTM and GANs-CNN models, possessing a tensor size of (9, 9), demonstrated enhanced performance in comparison to the LSTM and CNN models. The enhanced performance of the GANs-CNNLSTM architecture in comparison to the GANs-LSTM and GANs-CNN architectures can be attributed to its adeptness in effectively integrating spatial and temporal features. The E-GAN model has exhibited enhanced efficacy in comparison to its peer models. The reason for this phenomenon can be attributed to the inadequacy of conventional GANs in exploring the entire range of possible outcomes when exposed to noisy inputs. Consequently, their capacity to produce samples of superior quality is restricted.

The results obtained from GANs-ConvLSTM-n demonstrate that its accuracy is relatively lower when compared to that of GANs-LSTM and GANs-CNN. The effectiveness of basic generative adversarial networks (GANs) may be constrained in their capacity to comprehend the dynamic spatial-temporal correlation that exists in high-dimensional data, as opposed to their proficiency in low-dimensional data. This study contributes to the existing literature by integrating variational inference with generative adversarial networks (GANs) through the utilisation of the convolutional long short-term memory (ConvLSTM) framework. The GANs-ConvLSTM-s technique exhibits superior performance compared to alternative approaches due to its ability to effectively manage the intricate interplay between structured and stochastic characteristics inherent in spatio-temporal (ST) data. The empirical evidence indicates that E-GAN outperforms GANs-ConvLSTM-s, despite sharing identical architecture and parameter count, as the latter exhibits inferior performance.

This discovery demonstrates the efficacy of integrating external variables with two-tiered temporal correlation in forecasting future events. The proposed solution is denominated E-GAN. The findings indicate a noteworthy decline in performance, as even a pixel-wise comparison of two spatio-temporal (ST) maps may result in the loss of certain minor features. It is crucial to bear in mind that the deep generative adversarial network for feature selection (D-GANFS) fails to converge in the majority of instances.

In the context of urban applications that utilise spatio-temporal data, such as the prediction of taxi demand or crowd flow, a grid within the spatio-temporal map or a pixel within an image represents the frequency of requests for taxis or bikes within a specific geographic area over a designated period of time. The primary objective is to minimise the disparity between the data generated and the actual data. The proposed method employs the feature loss to demonstrate the significance of the selected pixel loss.

5.4.5 Analysis of External Factors Fusion

The analysis of E-GAN involves the integration of additional data into the ST correlation module. The process of integration enhances the effectiveness of predictions. The analysis focuses on the variables of point of interest (PoI), weekday/weekend (Dw), and weather data (W). The correlation between meteorological data and taxi trips is illustrated in Fig. 5.5.

The spatial arrangement of New York City's points of interest (PoIs) is depicted in Fig. 5.6. This research incorporates extraneous variables to investigate two inquiries. (i) Can external synergy yield favourable outcomes? (ii) Can the effectiveness of E-GAN be influenced by a particular external factor?

The proposed model, E-GAN-ExF-n, does not incorporate external factors. The proposed model, E-GAN-ExF-c, incorporates external factors through direct concatenation. The E-GAN-ExF-1 model incorporates a single point of interest (PoI) data as an external factor for fusion. The proposed model, E-GAN-ExF-2, incorporates the fusion of PoI and Dw. The proposed model, E-GAN-ExF-3/E-GAN, incorporates all external factors. The present document comprises the comprehensive rendition of our proposed model, namely E-GAN. Table 5.3 illustrates the influence of extrinsic variables on the performance of E-GAN. It has been observed that the complete iteration of the E-GAN model that we have developed exhibits superior performance in the majority of the evaluation metrics. Specifically, we provide a summary of two significant observations:

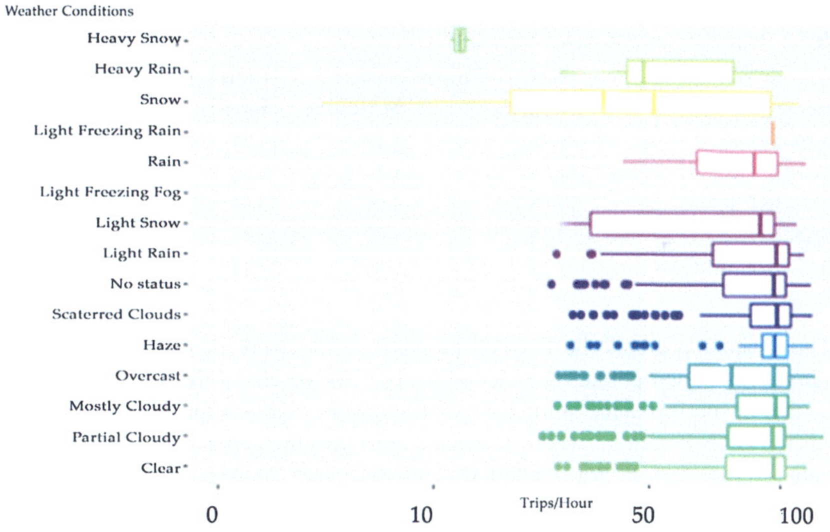

Fig. 5.5 Weather impact during taxi trips

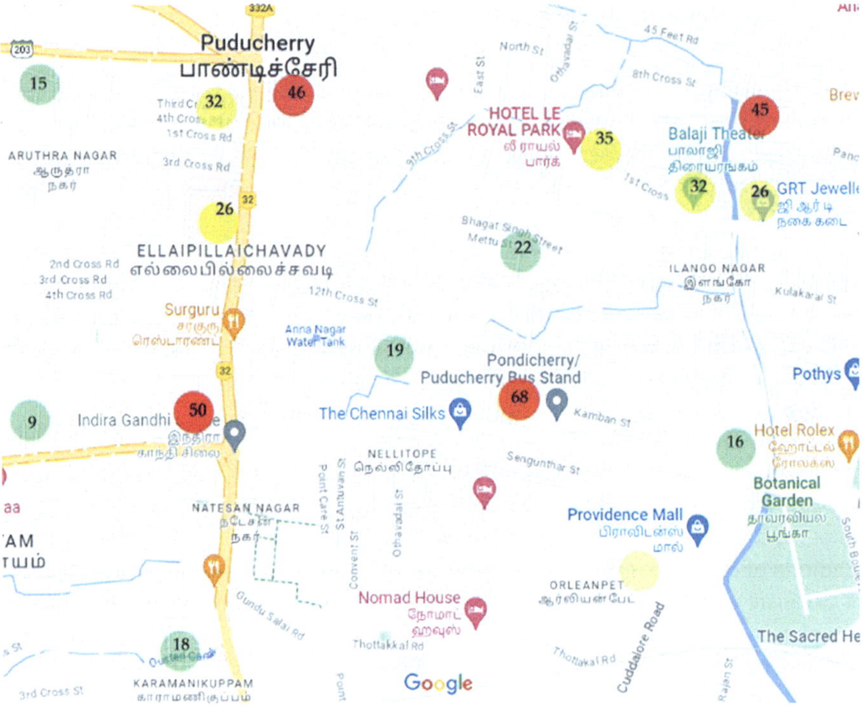

Fig. 5.6 PoI distribution in NYC area

Table 5.3 Impact of external factors on E-GAN

	NYC TLC trip record data		CitiBikeNYC dataset	
Models	RMSE	MAE	RMSE	MAE
E-GAN-ExF-n	1.42	1.19	4.21	2.09
E-GAN-ExF-c	1.54	1.09	4.32	1.54
E-GAN-ExF-1	1.65	1.06	3.98	1.78
E-GAN-ExF-2	1.56	1.02	3.45	1.11
E-GAN-ExF-3/E-GAN	1.21	0.97	3.11	1.02

This chapter presents the E-GAN predictive framework in its comprehensive iteration, which incorporates all external factors. The results of the analysis demonstrate that the performance of the E-GAN framework in this iteration surpasses that of other E-GAN iterations, where external factors are either concatenated directly or not all external factors are integrated. This discovery showcases the efficacy of external variables and their integration in augmenting the model's efficacy. The results suggest that there is a similar improvement in the predictive capabilities of the model for both E-GAN-ExF-c and E-GAN-ExF-n. The E-GAN-ExF-c model does not possess a clearly defined objective for the identification of correlations between disparate data sources (Fig. 5.7).

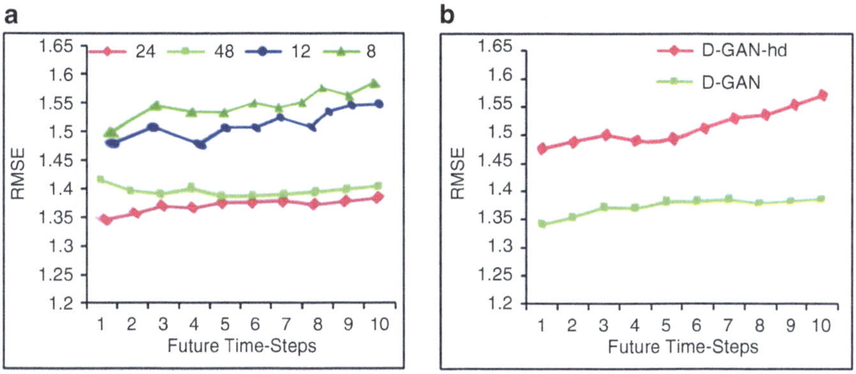

Fig. 5.7 (**a**, **b**) Impact on RMSE-based sequence length and two-level temporal correlation

Table 5.4 Impact of two-level temporal correlation on E-GAN

	NYC TLC trip record data		CitiBikeNYC dataset	
Models	RMSE	MAE	RMSE	MAE
E-GAN-V	3.39	3.16	6.18	4.06
E-GAN-hd	3.51	3.06	6.29	3.51
E-GAN	1.21	0.97	3.11	1.02

The results suggest that E-GAN demonstrates enhanced efficacy in comparison to E-GAN-ExF-1 and E-GAN-ExF-2. The EGAN model is proficient in acquiring knowledge of both spatio-temporal (ST) correlations and fundamental factors of variations through the modelling of the distribution of ST data. As a result, the incorporation of extraneous variables, such as point of interest (PoI) or day of the week (Dw), does not result in noteworthy enhancements. The improved ability to predict outcomes highlights the superiority of our fusion network.

5.4.6 Effect of Two-Level Temporal Correlation

The current investigation presents a novel approach, namely, the two-level temporal correlation methodology, for forecasting spatio-temporal data. The objective of this methodology is to incorporate a wider spectrum of temporal interdependencies. The present study aims to examine the impact of temporal correlation at two levels by evaluating the efficacy of our E-GAN model under varying data mapping scenarios.

E-GAN-hd: The model under consideration accepts a three-dimensional tensor as its input, which represents the spatial-temporal (ST) data of a given day. Each individual ST map contained in the tensor is representative of the ST data collected over a 1-h time period. Table 5.4 displays the impact of data mapping on the performance of E-GAN. It is observed that the fully realised iteration of the proposed

E-GAN model exhibits superior performance across a majority of evaluation metrics. Specifically, we provide a summary of two pivotal observations.

In this context, each subsequent action involves the anticipation of 24 frames. Initially, an analysis is conducted to investigate the impact of the sequence length of spatio-temporal (ST) maps on the efficacy of the model, as depicted. The results for this case are approximately similar in the CitiBikeNYC dataset, therefore they have been omitted. The optimal performance for a sequence length of 24 h is attained by E-GAN-hd. The root mean square error (RMSE) exhibited a degradation for the duration of 8 and 12 h. One possible explanation is that the model may have failed to capture temporal dependencies. As the length of the sequence increases to 48, the prediction error experiences a slight degradation at step 1, whereas in subsequent steps, the performance is roughly equivalent to that observed at 24 h. The rationale behind the challenge of calculating extended temporal dependency lies in the need to train a greater number of parameters, resulting in increased difficulty in the training process. The proposed work involves predicting 24 frames at a time for each subsequent step. The findings indicate that E-GAN outperforms E-GAN-hd. It is worth noting that the CitiBikeNYC dataset yielded comparable results in this scenario, but we have omitted them for brevity. The potential explanation is that the model can effectively capture nuanced temporal dependencies by utilising the two-level temporal correlation. The superior performance of E-GAN over E-GAN-hd provides evidence of E-GAN's ability to effectively capture intricate and extended temporal dependencies.

5.5 Summary

This chapter introduced a new deep stochastic enhanced adversarial-based network, denoted as E-GAN, for the purpose of ST prediction across multiple time intervals. The E-GAN model has been shown to be good at figuring out how implicitly spatial-temporal (ST) data is distributed. This lets correlations, underlying factors of variation, and the data's multimodality be modelled. The research aims to create an enhanced spatio-temporal correlation network model that captures data variation and multimodality. A fusion network with a precise objective function detects correlations between spatio-temporal (ST) data and external factors. Implicit distribution modelling with a modified stochastic adversarial network generates and infers ST data. Dual-level temporal correlation captures complex temporal dynamics. Through diverse data sources and easy modification, the E-GAN model can address novel spatio-temporal (ST) challenges. Two real datasets tested our demand prediction model. Our model predicts crowd flow to demonstrate its spatio-temporal forecasting capabilities across a variety of urban use cases. E-GAN outperforms traditional methods in both contexts. The present work offers novel perspectives on the modelling of intricate spatial and temporal connections and the concurrent capture of data variations. The model proposed by the authors presents a straightforward yet efficient approach to address the challenge of predicting spatio-temporal patterns in extensive continuous spaces across multiple modalities. The authors anticipate that this model will prove beneficial in domains characterised by high levels of uncertainty. In subsequent research, we will explore methods to enhance the current study by integrating various sources of spatio-temporal data, including but not limited to bus, bike, and truck data, mobile usage data, and mobility data, to facilitate collective prediction.

References

1. Esteva, Andre, Katherine Chou, Serena Yeung, Nikhil Naik, Ali Madani, Ali Mottaghi, Yun Liu, Eric Topol, Jeff Dean, and Richard Socher. "Deep learning-enabled medical computer vision." *NPJ digital medicine* 4, no. 1 (2021): 5.
2. Schranz, Melanie, Gianni A. Di Caro, Thomas Schmickl, Wilfried Elmenreich, Farshad Arvin, Ahmet Şekercioğlu, and Micha Sende. "Swarm intelligence and cyber-physical systems: concepts, challenges and future trends." *Swarm and Evolutionary Computation* 60 (2021): 100762.
3. Hamdy, Omar, Hanan Gaber, Mohamed S. Abdalzaher, and Mahmoud Elhadidy. "Identifying exposure of urban area to certain seismic hazard using machine learning and GIS: A case study of greater Cairo." *Sustainability* 14, no. 17 (2022): 10722.
4. Ali, Ahmad, Yanmin Zhu, and Muhammad Zakarya. "Exploiting dynamic spatio-temporal graph convolutional neural networks for citywide traffic flows prediction." *Neural networks* 145 (2022): 233-247.
5. Cao, Yibo, Lu Liu, and Yuhan Dong. "Convolutional Long Short-Term Memory Two-Dimensional Bidirectional Graph Convolutional Network for Taxi Demand Prediction." *Sustainability* 15, no. 10 (2023): 7903.
6. Rajmohan, R., M. Pavithra, T. Ananth Kumar, and P. Manjubala. "Exploration of deep RNN architectures: LSTM and gru in medical diagnostics of cardiovascular and neuro diseases." In *Handbook of Deep Learning in Biomedical Engineering and Health Informatics*, pp. 167-202. Apple Academic Press, 2021.
7. Orrù, Pier Francesco, Andrea Zoccheddu, Lorenzo Sassu, Carmine Mattia, Riccardo Cozza, and Simone Arena. "Machine learning approach using MLP and SVM algorithms for the fault prediction of a centrifugal pump in the oil and gas industry." *Sustainability* 12, no. 11 (2020): 4776.
8. Yang, Jianfeng, Ru Li, Liangchao Chen, Yuanhao Hu, and Zhan Dou. "Research on equipment corrosion diagnosis method and prediction model driven by data." *Process Safety and Environmental Protection* 158 (2022): 418-431.
9. Zhang, Dingcheng, Edward Stewart, Mani Entezami, Clive Roberts, and Dejie Yu. "Intelligent acoustic-based fault diagnosis of roller bearings using a deep graph convolutional network." *Measurement* 156 (2020): 107585.
10. Liu, Shaowei, Hongkai Jiang, Zhenghong Wu, and Xingqiu Li. "Data synthesis using deep feature enhanced generative adversarial networks for rolling bearing imbalanced fault diagnosis." *Mechanical Systems and Signal Processing* 163 (2022): 108139.
11. Suresh Kumar, K., and C. Helen Sulochana. "Local search five-element cycle optimized reLU-BiLSTM for multilingual aspect-based text classification." *Concurrency and Computation: Practice and Experience* 34, no. 28 (2022): e7374.
12. Vanichrujee, Ukrish, Teerayut Horanont, Wasan Pattara-atikom, Thanaruk Theeramunkong, and Takahiro Shinozaki. "Taxi demand prediction using ensemble model based on rnns and xgboost." In *2018 International Conference on Embedded Systems and Intelligent Technology & International Conference on Information and Communication Technology for Embedded Systems (ICESIT-ICICTES)*, pp. 1-6. IEEE, 2018.

13. Ge, Liang, Siyu Li, Yaqian Wang, Feng Chang, and Kunyan Wu. "Global spatial-temporal graph convolutional network for urban traffic speed prediction." *Applied Sciences* 10, no. 4 (2020): 1509.
14. Alonso-González, María J., Theo Liu, Oded Cats, Niels Van Oort, and Serge Hoogendoorn. "The potential of demand-responsive transport as a complement to public transport: An assessment framework and an empirical evaluation." *Transportation Research Record* 2672, no. 8 (2018): 879-889.
15. Ganegoda, Naleen Chaminda, Karunia Putra Wijaya, Joseph Páez Chávez, Dipo Aldila, KKW Hasitha Erandi, and Miracle Amadi. "Reassessment of contact restrictions and testing campaigns against COVID-19 via spatio-temporal modeling." *Nonlinear Dynamics* 107, no. 3 (2022): 3085-3109.
16. Sundaresan, S., K. Suresh, V. Kishore, and A. Jayakumar. "Insight Into Various Algorithms For Medical Image Analyzes Using Convolutional Neural Networks (Deep Learning)." In *Handbook of Deep Learning in Biomedical Engineering and Health Informatics*, pp. 137-164. Apple Academic Press, 2021.
17. Saxena, Divya, and Jiannong Cao. "Multimodal Spatio-Temporal Prediction with Stochastic Adversarial Networks." *ACM Transactions on Intelligent Systems and Technology (TIST)* 13, no. 2 (2022): 1-23.
18. Aburas, Maher Milad, Yuek Ming Ho, Mohammad Firuz Ramli, and Zulfa Hanan Ash'aari. "The simulation and prediction of spatio-temporal urban growth trends using cellular automata models: A review." *International Journal of Applied Earth Observation and Geoinformation* 52 (2016): 380-389.
19. Martínez-Arellano, Giovanna, German Terrazas, and Svetan Ratchev. "Tool wear classification using time series imaging and deep learning." *The International Journal of Advanced Manufacturing Technology* 104 (2019): 3647-3662.
20. Chattopadhyay, Ashesh, Pedram Hassanzadeh, and Saba Pasha. "Predicting clustered weather patterns: A test case for applications of convolutional neural networks to spatio-temporal climate data." *Scientific reports* 10, no. 1 (2020): 1317.
21. Fang, Zheng, Qingqing Long, Guojie Song, and Kunqing Xie. "Spatial-temporal graph ode networks for traffic flow forecasting." In *Proceedings of the 27th ACM SIGKDD conference on knowledge discovery & data mining*, pp. 364-373. 2021.

Chapter 6
Spatiotemporal Object Detection and Activity Recognition

Vimal Kumar, Shobhit Jain, and David Lillis

Abstract Spatiotemporal object detection and activity recognition are essential components in the advancement of computer vision, with broad applications spanning surveillance, autonomous driving, and smart stores. This chapter offers a comprehensive overview of the techniques and applications associated with these concepts. Beginning with an introduction to the fundamental principles of object detection and activity recognition, we discuss the challenges and limitations posed by existing methods. The chapter progresses to explore spatiotemporal object detection and activity recognition, which entails capturing spatial and temporal information of moving objects in video data. A hierarchical model for spatiotemporal object detection and activity recognition is proposed, designed to maintain spatial and temporal connectivity across frames. Additionally, the chapter outlines various metrics for evaluating the performance of object detection and activity recognition models, ensuring their accuracy and effectiveness in real-world applications. Finally, we underscore the significance of spatiotemporal object detection and activity recognition in diverse fields such as surveillance, autonomous driving, and smart stores, emphasizing the potential for further research and development in these areas. In summary, this chapter provides a thorough examination of spatiotemporal object detection and activity recognition, from the foundational concepts to the latest techniques and applications. By presenting a hierarchical model and performance evaluation metrics, the chapter serves as a valuable resource for researchers and practitioners seeking to harness the power of computer vision in a variety of domains.

Keywords Spatiotemporal data · Object detection · Activity recognition Challenges and limitations

V. Kumar (✉) · D. Lillis
University College Dublin, Dublin, Ireland
e-mail: vimal.kumar@ucd.ie; david.lillis@ucd.ie

S. Jain
The University of Texas at Dallas, Richardson, TX, USA

J. A et al. (eds.), *Spatiotemporal Data Analytics and Modeling*, Big Data Management, https://doi.org/10.1007/978-981-99-9651-3_6

6.1 Introduction

Spatiotemporal data analytics is an interdisciplinary field that uses computational methods and algorithms to examine data that changes over space and time. With advancements in sensing technologies and a rapid increase in data volume, the significance of spatiotemporal analytics has grown exponentially, finding extensive applications in fields such as environmental monitoring, urban planning, transportation, health surveillance, and increasingly in intelligent systems like autonomous vehicles and smart stores [1]. Central to this interdisciplinary domain is spatiotemporal data that captures information across both spatial and temporal dimensions. Its importance has amplified with technological advancements and the widespread use of surveillance systems, wearable devices, and video platforms. One pivotal application of spatiotemporal data is in object detection and activity recognition in videos, which involves identifying, localizing, and classifying activities performed by specific objects in video sequences [2].

Object detection, an integral component of any video understanding system, seeks to locate and classify objects of interest within video frames. Despite challenges posed by variations in object appearance, occlusions, and changes in camera viewpoint, this process has witnessed significant improvements due to the advent of deep learning techniques [3–5]. After object detection, activity recognition comes into play, focusing on identifying the actions or behaviors exhibited by the detected objects over a sequence of frames. Activity recognition finds applications across diverse contexts, including sports, surveillance, health care, and wildlife tracking, despite the inherent ambiguity and variability in human actions [2, 4, 6, 7].

This chapter seeks to provide a comprehensive overview of the techniques for spatiotemporal object detection and activity recognition. We will discuss the evolution of these methodologies, current challenges and limitations, and state-of-the-art techniques. Moreover, the chapter will outline various performance evaluation metrics crucial for ensuring model accuracy and effectiveness. It will also explore the application of spatiotemporal object detection and activity recognition in diverse domains, including surveillance, autonomous driving, and smart stores, thereby showcasing current research and the potential for future advancements.

Following this introduction, Sect. 6.2 will delve into the history and basic techniques of object detection and activity recognition. Section 6.3 will address the challenges and limitations in this field, illustrated with examples. Section 6.4 will focus on the relevance of spatiotemporal data to object detection and activity recognition. Section 6.5 will offer a detailed discussion of the hierarchical model, while Sect. 6.6 will cover various performance evaluation metrics with comparison across different models. Lastly, Sect. 6.7 will discuss applications and potential future developments in the field.

In conclusion, this chapter aims to serve as a comprehensive guide to spatiotemporal object detection and activity recognition. It intends to offer a broad understanding of the concepts and techniques, proving invaluable for both novice readers and experienced researchers keen on exploring the dynamic world of video-based understanding systems. By summarizing key points, exploring potential impact on computer vision, and potential for future research, it aims to foster further exploration and innovation in this exciting field.

6.2 Fundamentals of Object Detection and Activity Recognition

To comprehend the complexity and breadth of object detection and activity recognition, we begin by exploring their historical progression, proceed to deconstruct the underlying algorithms, and finally delve into their importance and diverse applications across various industries.

Object detection, a cornerstone of computer vision systems, can trace its evolution back to the incipient stages of image processing. In the beginning, the goal was relatively straightforward: to identify and categorize objects within static images [8]. However, with the advent and proliferation of video technologies, the need to decipher moving visual content became paramount, propelling object detection to a whole new realm of challenges and possibilities. Early object detection methods primarily focused on discerning simple, distinct features such as edges or corners, that were considered as characteristic signatures of objects within an image. The emergence of machine learning and, more specifically, convolutional neural networks (CNNs) marked a significant breakthrough, empowering modern techniques to accurately identify and classify multiple objects within a single frame [9].

Activity recognition, an advancement of object detection, seeks to comprehend the dynamic behavior of detected objects within video sequences. This process of deciphering the semantics of motion necessitates not only recognizing the individual actions performed by objects but also understanding the context and temporal sequencing of these actions [10, 11].

At a rudimentary level, object detection and activity recognition operate on a sequential pipeline comprising preprocessing of data, feature extraction, and finally, classification or recognition (Fig. 6.1). The initial preprocessing stage involves

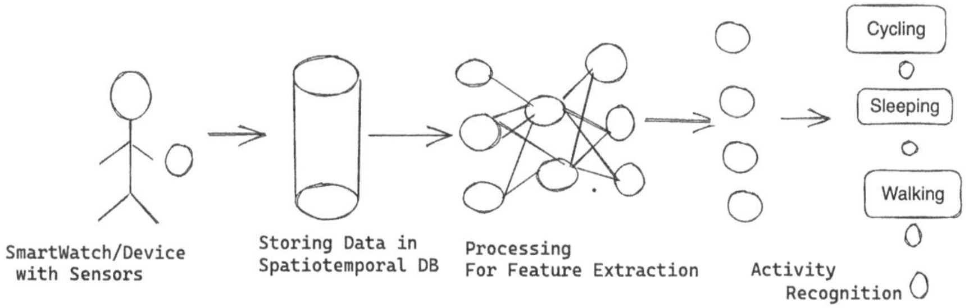

Fig. 6.1 The pipeline of object detection and activity recognition

filtering noise and normalizing the input data to a standard format. Following this, the feature extraction phase seeks to distinguish key characteristics that help differentiate various objects or activities. In the context of object detection, these features could be geometric attributes such as shape, color intensity, or texture gradients. Conversely, for activity recognition, spatiotemporal features like motion trajectories, sequential object interactions, or action primitives form the crux [8]. The final stage of the pipeline involves classification or recognition, which employs machine learning algorithms to segregate the extracted features into distinct classes or activities.

Object detection and activity recognition techniques have demonstrated immense utility and versatility across numerous sectors. For instance, in surveillance systems, these techniques enable detection of anomalies or intrusions, effectively enhancing security measures [9]. In health care, object detection and activity recognition can contribute toward patient behavior monitoring and assist in intricate surgical procedures [6]. In the realm of autonomous vehicles, object detection facilitates the recognition of pedestrians, vehicles, and potential obstructions, whereas activity recognition aids in interpreting their movement patterns to predict future trajectories. Even in sports analytics, these techniques can augment player performance evaluation and strategizing by comprehending player movements and actions [2].

To summarize, the rapid evolution and convergence of object detection and activity recognition techniques have been instrumental in the development of sophisticated systems capable of understanding and interpreting dynamic visual content. The integration of these techniques with cutting-edge machine learning algorithms has not only expanded the boundaries of possibilities but also paved the way for novel challenges in the field of spatiotemporal data analytics (Table 6.1). In the ensuing section (Sect. 6.3), we shall delve deeper into the specific challenges and potential roadblocks encountered in contemporary techniques of object detection and activity recognition, paving the way for discussion on future research directions.

Table 6.1 A summary highlighting diverse applications of object detection and activity recognition across various sectors

Sector	Applications of object detection and activity recognition
Surveillance and security	(a) Enable detection of anomalies and intrusions (b) Monitor estates in real time (c) AI powered video analytics to find anomalies in a recorded video footage
Sports	(a) Assist referee in match decisions (b) Evaluate player's performance
Health care	(a) Patient behavior monitoring (b) Assist in critical surgeries
Retail	(a) Contactless checkout by creating a virtual shopping cart using AI (b) Inventory management to monitor stocks and its arrangements for easy access (c) Foot traffic analysis to perform crowd management (d) Informing employees for customer assistance when there is a need
Industries	(a) Monitoring the use of personal protective equipment (PPE) kits by workers (b) Product assembly to enable automated production in industries (c) Defect detection to maintain quality standards (d) Productivity improvement by monitoring workers activity
Transportation and automobile	(a) Driverless/autonomous vehicles (b) Collision detection (c) People counting for safety and planning (d) Parking occupancy monitoring (e) In-cabin monitoring for safety and medical emergencies (f) Road maintenance analysis

6.3 Challenges and Limitations in Object Detection and Activity Recognition

The promising developments in object detection and activity recognition have substantially aided our comprehension and interpretation of visual data. However, there exist numerous formidable challenges and limitations associated with these methodologies that need to be addressed.

One of the foremost challenges in object detection arises from substantial variations in scale, pose, and illumination of objects that are frequently encountered in real-world scenarios [12]. Objects of different sizes, orientations, and lighting conditions demand different levels of discernment, often putting a strain on

Fig. 6.2 Instance of occlusion in a crowded scene [14]

conventional detection techniques. Furthermore, the instances of overlapping objects and occlusion also compound the problem of accurate object detection. Occlusion, where objects are partially or completely hidden behind other objects, often leads to detection failures [13]. Figure 6.2 can depict an instance of occlusion in a crowded scene, demonstrating the complexity of accurate detection.

Another recurring issue in object detection is the detection of small, thin, or tiny objects [1]. The effective detection and classification of such objects remain challenging due to their poor representation in the feature maps. Traditional object detectors often fail to capture the nuanced details of such objects, leading to reduced detection performance.

Alongside these, the presence of background clutter and highly similar objects also pose significant difficulties [5]. Distinguishing objects from their surroundings or differentiating between similar-looking objects requires the implementation of highly sophisticated feature extraction and classification mechanisms.

Turning to activity recognition, the challenges grow multifold due to the added temporal dimension. Activities are inherently temporal in nature and can exhibit significant variability in terms of duration, speed, and sequence [13]. The need to comprehend the context of activities, including the recognition of interactions between multiple entities or the interpretation of multi-activity scenes, poses a significant challenge [7]. Understanding these interactions is critical in many application scenarios, such as surveillance and elderly care, but it is still a challenging task for current systems.

Moreover, intra-class variability, where the same activity is performed differently by different individuals or by the same individual at different times, is another prominent issue in activity recognition [3]. Different individuals have unique styles or habits, and these variations can drastically affect the recognition results. A notable case that brings these challenges to the forefront is in the field of sports video behavior recognition [2]. Identifying individual players' actions, interpreting team strategies, or recognizing significant events such as goals in soccer or aces in tennis necessitates sophisticated object detection and activity recognition methodologies. These tasks become exceedingly complex due to fast player movements, occlusion scenarios, and constant changes in the field of view.

In conclusion, while object detection and activity recognition technologies have made considerable strides, significant challenges and limitations persist. Further

Table 6.2 Encapsulates the main challenges and limitations in object detection and activity recognition

Serial number	Current challenges and limitations	Description
1	Interactions between multiple entities simultaneously	It will be more complex when a scene contains multiple objects and their activities during interactions.
2	Multi-activity scenarios	When there are two objects performing different actions, the complexity of identifying it through different frames becomes challenging.
3	Intra-class variability and identification	In a people monitoring application, there are different styles of walking by each individual and it creates ambiguity in identification of actions.
4	Sports video behavior recognition	For detecting fouls in a sport, sports video will be really helpful, but, since players are already in motion, it will be a major challenge to recognize.

research and technological advancements are required to address these issues and enhance the robustness and adaptability of these systems to effectively handle real-world variability and complexity (Table 6.2).

6.4 Spatiotemporal Object Detection and Activity Recognition

Spatiotemporal object detection and activity recognition concern the understanding of both spatial and temporal dynamics within video data, crucial for a comprehensive representation of events and activities in the real world. Here, the spatial aspect refers to the recognition of objects in the video frames, while the temporal aspect refers to the understanding of these objects' behaviors and movements over time.

The intrinsic value of incorporating spatiotemporal information within video data analysis lies in its potential to capture more nuanced and sophisticated understandings of complex events. For example, in sports analytics, not only the recognition of individual players and their actions (spatial) but also the pattern and sequence of these actions (temporal) are of paramount importance for strategy analysis [2]. Figure 6.3 can demonstrate the process of spatiotemporal object detection and activity recognition within a sports game scenario.

In another context, an application of spatiotemporal recognition is visible in surgical procedures. Here, the recognition of surgical instruments (spatial) and their usage over time (temporal) can aid in monitoring and evaluating surgical procedures [6]. Despite its significant potential, spatiotemporal object detection and activity

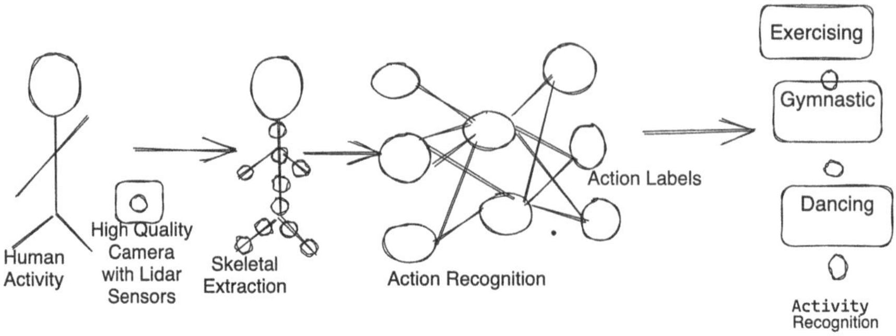

Fig. 6.3 Spatiotemporal object detection and activity recognition in a sports game

recognition present complex challenges. One of these challenges involves the extraction of useful spatiotemporal features from raw video data, which is a computationally expensive process requiring efficient and effective algorithms [11]. Moreover, the successful segmentation of foreground objects, which is crucial for activity semantics understanding, remains a difficult task due to varying lighting conditions, occlusions, and complex background activities [15]. Another key challenge involves the accurate detection and recognition of activities in real time. Many applications, such as surveillance systems and elderly care service robots, require instant responses, making the temporal aspect of activity detection significantly important [3].

Recognizing activities that involve multiple interacting objects is also a formidable challenge. Multiple objects can perform interdependent activities, resulting in a complex scenario that demands not only the recognition of individual activities but also an understanding of the interactions between them [7]. Recent advancements have addressed some of these challenges using sophisticated methodologies. For instance, using spatiotemporal depth cuboid similarity features extracted from depth camera data has shown promising results in activity recognition [16]. Further, employing the idea of spatiotemporal attention-based LSTM networks has achieved a significant breakthrough in 3D action recognition and detection [17].

In summary, spatiotemporal object detection and activity recognition is a burgeoning area of research with the potential to drastically improve our understanding and interpretation of complex real-world events. Yet, it also comes with its own set of challenges that necessitate further technological advancements and innovative research.

6.5 Model for Spatiotemporal Object Detection and Activity Recognition

This section focuses on the hierarchical model for spatiotemporal object detection and activity recognition. The model maintains both spatial and temporal connectivity across frames, a crucial feature for robust and efficient video understanding.

6.5.1 Hierarchical Model

The hierarchical model in this research is inspired by the concept of maintaining spatial and temporal connectivity across frames in videos. It adopts a twofold approach for activity recognition, which involves spatial object detection followed by temporal activity recognition. The object detection stage involves identifying key objects in each frame, and then the activity recognition stage analyzes the temporal evolution of these objects over several frames. The entire process can be viewed as a dynamic pipeline (as shown in Fig. 6.4), with object detection feeding into activity recognition.

Object detection relies on a robust foreground object segmentation mechanism, where key objects of interest are identified and segmented from the background [5, 15]. This approach is often employed in sports video behavior recognition, where key objects like players and the ball need to be identified for effective activity analysis [2, 18].

The model incorporates a multiscale spatiotemporal context approach, similar to those found in related work [11, 17]. It maintains the spatiotemporal continuity across frames by leveraging advanced computer vision techniques like Hough forests [19], LSTM networks with spatiotemporal attention [17], and depth cuboid similarity features [16]. Activity recognition, on the other hand, utilizes a combination of advanced temporal augmentation strategies and machine learning techniques [6]. This stage of the model emphasizes on the temporal evolution of key objects, recognizing patterns that signify specific activities. This aspect is important in applications like surgical activity recognition, where it is crucial to understand the temporal evolution of activities from video data [6].

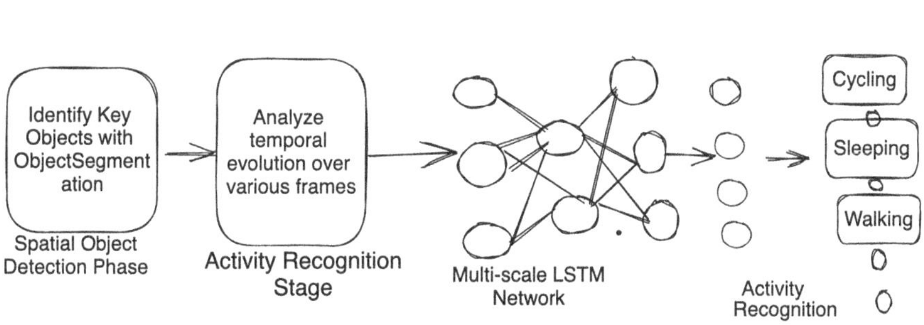

Fig. 6.4 The Hierarchical model for spatiotemporal object detection and activity recognition

Table 6.3 Comparison of different techniques for maintaining spatiotemporal connectivity

S. No.	Techniques	Merits	Demerits
1	Hierarchical framework [10]	Encodes contextual information at different levels—Point, intra-trajectory, and intertrajectory.	Relative motion between objects can reduce the performance of action recognition.
2	Human-object interaction model [20]	(i) Human action is predicted using human-object interaction as spatiotemporal features. (ii) Computationally efficient and can be used in real-time processing.	Convolution networks will not be able to encode semantic information into the spatiotemporal features.
3	Self-supervised framework [12]	(i) To reduce computational complexity, it uses patches extracted from video frames divided. (ii) To reduce memory constraints, this approach uses a permutation strategy instead of a random strategy.	No significant demerits identified.
4	Semantic representation technique [21]	(i) A graph driven approach to detect visual patterns. (ii) It has faster search time.	Event patterns at different hierarchies are not handled.
5	Cascaded region proposal and detection (CRPAD) framework [9]	This approach identifies activities at a fine-grain level.	Temporal modeling is not used in this approach.

6.5.2 *Spatial and Temporal Connectivity*

This model's unique feature is its emphasis on maintaining spatial and temporal connectivity across frames, thereby creating a holistic understanding of the scene's dynamic evolution. This connectivity is maintained via an efficient spatial-temporal graph, similar to what was proposed by Sun et al. [10]. This graph-based approach enables the model to establish object-object and object-scene relations, which contribute to a deeper understanding of the activity being performed [20]. This concept of connectivity is particularly significant in complex video environments, such as those found in surveillance or sports videos, where the context and multiscale motion awareness play a significant role in activity recognition [7, 9]. Additionally, the model incorporates advanced techniques for maintaining spatiotemporal connectivity. These include knowledge graphs for spatiotemporal event pattern matching [21], optimization models for human activity recognition inspired by information on human-object interaction [20], and the application of unsupervised learning of spatiotemporal context for video action recognition [12] (Table 6.3).

6.6 Performance Analysis of Object Detection and Activity Detection

In this section, we present and discuss various metrics used to evaluate the performance of object detection and activity recognition models. These metrics are paramount for ensuring model accuracy and effectiveness. In complex video understanding tasks, the effectiveness of a model can be evaluated by how well it performs in spatial object detection and temporal activity recognition, quantified using specific metrics [2, 15].

6.6.1 Analysis for Object Detection

Precision (P): This metric measures the proportion of correctly detected objects among all detections made by the model. It is calculated as:

$$P = \frac{TP}{TP + FP}$$

where TP is the number of true positives, and FP is the number of false positives [4].

Recall (R): This metric measures the proportion of correctly detected objects among all the ground truth objects. It is calculated as:

$$R = \frac{TP}{TP + FN}$$

where FN is the number of false negatives [7].

F1 score: This is the harmonic mean of precision and recall, providing a single metric that balances both. It is calculated as

$$F1 = 2 * \frac{P * R}{P + R}$$

It is particularly useful when the cost of false positives and false negatives are significantly different [18].

Intersection over union (IoU): This metric measures the overlap between the predicted bounding box and the ground truth bounding box. A high IoU score means the prediction closely aligns with the ground truth. It is calculated as

$$\text{IoU} = \frac{\text{Area of overlap}}{\text{Area of union}}$$

IoU is an essential metric for object detection tasks as it quantifies the goodness of fit of the predicted bounding box [3].

6.6.2 Analysis for Activity Recognition

For activity recognition, the following metrics are commonly used:

Classification accuracy: This is the proportion of correctly classified activities among all activities. It is calculated as

$$\text{Accuracy} = \frac{TP + TN}{TP + FP + FN + TN}$$

where TN is the number of true negatives [20].

Mean average precision (mAP): This is the average precision at different recall levels. It calculates the area under the precision-recall curve and provides a single measure that balances precision and recall over all classes [6].

Confusion matrix: Although not a single metric, a confusion matrix gives a comprehensive view of how well the model performs across all classes of activities. It allows researchers to see the number of false positives and false negatives for each class, providing more insight into the model's performance [8] (Table 6.4).

As an example, consider the use case of a model detecting and recognizing activities in sports videos, such as tennis [18]. Here, the precise detection of players (object detection) and the accurate recognition of their activities (activity recognition) such as "serving," "forehand," "backhand," etc. are crucial. These evaluation metrics can provide a comprehensive performance assessment.

By carefully considering these metrics during the development and evaluation of models, researchers can ensure a thorough and rigorous performance evaluation. A high-performing model should demonstrate satisfactory performance in these metrics across multiple diverse datasets.

Table 6.4 Example of a confusion matrix for activity recognition [22]

True activity (down)/ predicted activity (right)	Boxing	HandClapping	HandWaving	Jogging	Running	Walking
Boxing	0.98	0.02	0	0	0	0
HandClapping	0.09	0.91	0	0	0	0
HandWaving	0.03	0.09	0.88	0	0	0
Jogging	0	0	0	0.83	0.10	0.07
Running	0	0	0	0.08	0.88	0.04
Walking	0	0	0	0.02	0	0.98

6.7 Applications of Spatiotemporal Object Detection and Activity Recognition

The fusion of spatiotemporal object detection and activity recognition carries substantial implications for a multitude of research fields, given its capacity to automate the understanding of object interactions and activities within a spatial and temporal context [13, 17]. The following section discusses these implications in the context of surveillance systems, autonomous vehicles, smart retail, sports analysis, and elderly care, demonstrating the breadth of applicability and potential avenues for future research [23].

6.7.1 Surveillance in Urban Environments

Spatiotemporal activity recognition enhances surveillance systems by enabling predictive and anomaly detection capabilities [10]. Techniques that incorporate context and multiscale motion awareness can facilitate advanced behavioral analysis, an area of research with promising implications for smart city management and public safety [7, 21, 24, 25]. Future research might consider how to fine-tune these systems for specific urban environments or unique public safety challenges (Fig. 6.5).

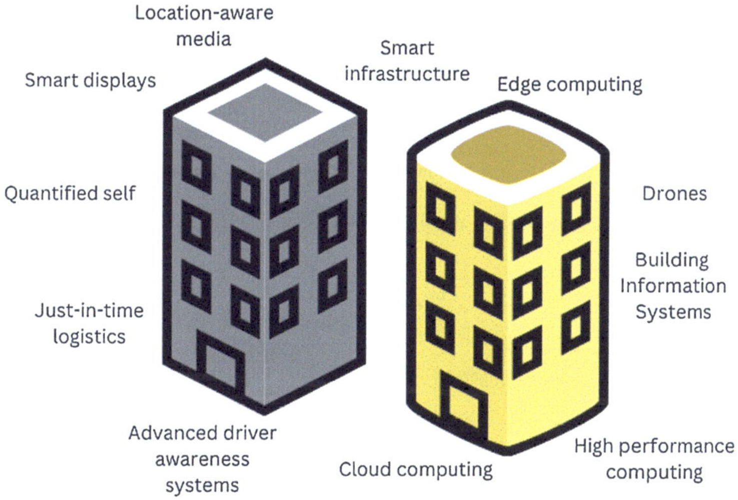

Fig. 6.5 Diagram depicting application of spatiotemporal object detection and activity recognition in urban surveillance

6.7.2 Autonomous Driving

Within the realm of autonomous driving, understanding dynamic traffic scenarios is critical, and spatiotemporal object detection offers a solution [10]. Research should continue to refine these models, ensuring they accurately recognize vehicles, pedestrians, and evolving traffic conditions. The potential for developing even more sophisticated models, capable of predicting and responding to unpredictable road scenarios, remains a compelling direction for further study (Fig. 6.6).

6.7.3 Smart Stores

The retail sector can utilize these technologies to gain detailed insights into customer behaviors and interactions [15, 20]. By analyzing customer-object interactions, retailers can optimize store layout and sales strategies. This opens a new area of research focused on refining and expanding the capabilities of these systems to provide more nuanced consumer behavior insights (Fig. 6.7).

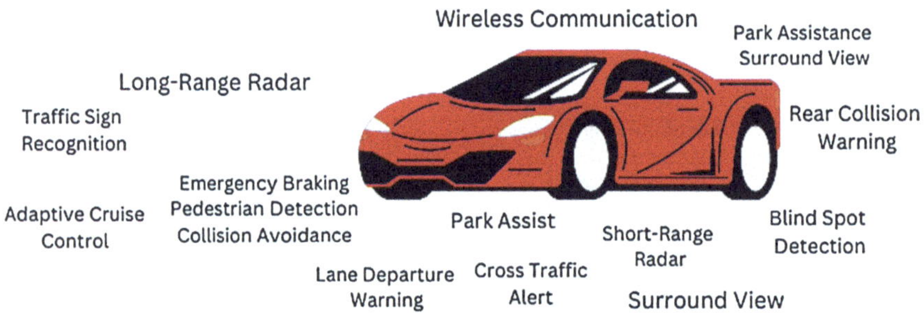

Fig. 6.6 Diagram showing the use of these technologies in autonomous vehicles

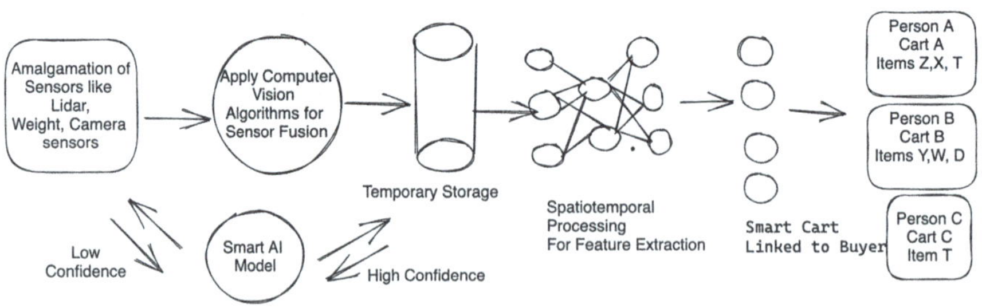

Fig. 6.7 Illustration of the application of these techniques in smart retail

6.7.4 *Sports Analysis*

The sports industry increasingly employs spatiotemporal object detection and activity recognition for performance improvement and broadcasting [2, 18]. This application provides a rich area for exploration, particularly in the development of sport-specific algorithms or models capable of analyzing complex, rapid movements.

6.7.5 *Elderly Care*

These technologies offer innovative solutions for elderly care, as evidenced by Wang et al.'s [3] temporal action detection model. Future research could investigate how to further tailor these systems for individual care needs, exploring their potential in tracking health indicators, or predicting and preventing accidents.

Case Study: Wildlife Monitoring
A case study by Clapham et al. [4] displays the potential of these technologies in wildlife monitoring. They developed a deep learning model capable of identifying individual brown bears, providing invaluable data for conservation efforts. This work mirrors advancements in human activity analysis [8, 26], indicating potential for future research in animal behavior studies and conservation efforts.

As research continues to push the boundaries of these technologies, novel approaches such as TRandAugment [6], and the usage of Hough forests for action recognition [19], illuminate potential paths of exploration. By considering these findings, researchers can identify areas for advancement, fostering innovation in the application of spatiotemporal object detection and activity recognition across various sectors.

6.8 Summary

In this chapter, we have explored the fascinating field of spatiotemporal object detection and activity recognition, investigating its concepts, techniques, applications, and the significance it holds for computer vision research. Let us summarize the key findings and highlight the potential for future research.

Beginning with the fundamentals in Sect. 6.2, we emphasized the importance of spatiotemporal analysis in understanding object interactions and activities within dynamic environments. We discussed the challenges posed by temporal variations and the need for effective models that maintain spatial and temporal connectivity across frames.

Moving forward, Sect. 6.3 focused on spatiotemporal object detection, where we examined various techniques, including region-based methods, motion cues, and two-stream networks. These approaches enable the detection of objects and their trajectories over time, contributing to improved scene understanding and object recognition accuracy.

Section 6.4 delved into activity recognition, where we explored the techniques used to recognize and categorize human activities in videos. We discussed the use of local spatiotemporal patterns, context-awareness, and multiscale motion awareness to enhance activity understanding. These approaches have shown promising results in accurately recognizing and interpreting human actions in dynamic scenes.

In Sect. 6.5, we introduced the concept of hierarchical models for spatiotemporal object detection and activity recognition. These models leverage multi-level representations and preserve contextual information across different scales, leading to improved accuracy and effectiveness in complex scenes.

Section 6.6 delved into performance evaluation metrics, highlighting their importance in assessing the performance of object detection and activity recognition models. We discussed various metrics used to evaluate the accuracy, precision, recall, and F1 score of these models, providing researchers with tools to compare and benchmark their approaches effectively.

Furthermore, in Sect. 6.7, we explored the diverse applications of spatiotemporal object detection and activity recognition, including surveillance systems, autonomous driving, and smart stores. These applications have witnessed significant advancements in situational awareness, decision-making, and safety, showcasing the transformative impact of spatiotemporal analysis in various domains.

Considering these key points, it is evident that spatiotemporal object detection and activity recognition play a pivotal role in computer vision research. They provide the means to comprehend complex scenes, identify objects, track their interactions, and infer activities in a spatial and temporal context.

Looking ahead, the potential for further advancements in this field is vast. Researchers can focus on developing more robust and efficient models that can handle real-world challenges such as occlusions, scale variations, and complex environmental conditions. Exploring novel applications, such as fine-grained activity detection or behavior prediction in specific domains, can lead to significant contributions. Moreover, there is a growing need to address interpretability and explainability in spatiotemporal models to ensure transparency and build trust with end-users.

In conclusion, spatiotemporal object detection and activity recognition have emerged as vital components of computer vision research, enabling us to extract meaningful information from dynamic scenes. Their impact extends beyond academia, reaching various industries and domains. By fostering collaboration, sharing knowledge, and embracing the challenges, we can drive innovation, refine techniques, and unlock the full potential of spatiotemporal analysis.

References

1. Pei, T., Huang, Q., Wang, X., Chen, X., Liu, Y., Song, C., … Zhou, C. (2021). Big geodata aggregation: Connotation, classification, and framework. National Remote Sensing Bulletin, 25(11), 2153–2162. doi: https://doi.org/10.11834/jrs.20210480
2. Liu, Y., & Jing, H. (2022). A Sports Video Behavior Recognition Using Local Spatiotemporal Patterns. Mobile Information Systems, 2022. doi: https://doi.org/10.1155/2022/4805993
3. Wang, K., Li, X., Yang, J., Wu, J., & Li, R. (2021). Temporal action detection based on two-stream You Only Look Once network for elderly care service robot. International Journal of Advanced Robotic Systems, 18(4). doi: https://doi.org/10.1177/17298814211038342
4. Clapham, M., Miller, E., Nguyen, M., & Darimont, C. T. (2020). Automated facial recognition for wildlife that lack unique markings: A deep learning approach for brown bears. Ecology and Evolution, 10(23), 12883–12892. doi: https://doi.org/10.1002/ece3.6840
5. Akilan, T. (2018). Video foreground localization from traditional methods to deep learning (Doctoral dissertation, University of Windsor (Canada)).
6. Ramesh, S., Dall'Alba, D., Gonzalez, C., Yu, T., Mascagni, P., Mutter, D., Padoy, N. (2023). TRandAugment: temporal random augmentation strategy for surgical activity recognition from videos. International Journal of Computer Assisted Radiology and Surgery. doi: https://doi.org/10.1007/s11548-023-02864-8
7. Cardoso, D. B., Campos, L. C. B., & Nascimento, E. R. (2022). An Action Recognition Approach with Context and Multiscale Motion Awareness. In Proceedings - 2022 35th Conference on Graphics, Patterns, and Images, SIBGRAPI 2022 (pp. 73–78). Institute of Electrical and Electronics Engineers Inc. doi: https://doi.org/10.1109/SIBGRAPI55357.2022.9991807
8. SankaranNampoothiri, S., & Anoop BK (2014). Review on Vision based Human Activity Analysis. International Journal of Computer Applications, 99(2), 9–14. doi: https://doi.org/10.5120/17343-6240
9. Aakur, S., Sawyer, D., Balazia, M., & Sarkar, S. (2020). An examination of proposal-based approaches to fine-grained activity detection in untrimmed surveillance videos. In 2018 TREC Video Retrieval Evaluation, TRECVID 2018. National Institute of Standards and Technology (NIST).
10. Sun, J., Wu, X., Yan, S., Cheong, L. F., Chua, T. S., & Li, J. (2009). Hierarchical spatio-temporal context modeling for action recognition. In 2009 IEEE Conference on Computer Vision and Pattern Recognition, CVPR 2009 (pp. 2004–2011). IEEE Computer Society. doi: https://doi.org/10.1109/CVPRW.2009.5206721
11. Wang, J., Chen, Z., & Wu, Y. (2011, June). Action recognition with multiscale spatio-temporal contexts. In CVPR 2011 (pp. 3185-3192). IEEE.
12. Ahsan, U., Madhok, R., & Essa, I. (2019, January). Video jigsaw: Unsupervised learning of spatiotemporal context for video action recognition. In 2019 IEEE Winter Conference on Applications of Computer Vision (WACV) (pp. 179-189). IEEE.
13. Liu, L., Shao, L., Li, X., & Lu, K. (2016). Learning spatio-temporal representations for action recognition: A genetic programming approach. IEEE Transactions on Cybernetics, 46(1), 158–170. doi: https://doi.org/10.1109/TCYB.2015.2399172.
14. Haroon Idrees, Khurram Soomro and Mubarak Shah, Detecting Humans in Dense Crowds using Locally-Consistent Scale Prior and Global Occlusion Reasoning, Pattern Analysis and Machine Intelligence (PAMI), IEEE Transactions, 2015.
15. Yu, T. W., Sarwar, M. A., Daraghmi, Y. A., Cheng, S. H., Ik, T. U., & Li, Y. L. (2022). Spatiotemporal Activity Semantics Understanding Based on Foreground Object Segmentation: iCounter Scenario. IEEE Access, 10, 57748–57758. doi: https://doi.org/10.1109/ACCESS.2022.3178609

16. Xia, L., & Aggarwal, J. K. (2013). Spatio-temporal depth cuboid similarity feature for activity recognition using depth camera. In Proceedings of the IEEE conference on computer vision and pattern recognition (pp. 2834-2841).
17. Song, S., Lan, C., Xing, J., Zeng, W., & Liu, J. (2018). Spatio-Temporal Attention-Based LSTM Networks for 3D Action Recognition and Detection. IEEE Transactions on Image Processing, 27(7), 3459–3471. doi: https://doi.org/10.1109/TIP.2018.2818328
18. Kanimozhi, S., Mala, T., Kaviya, A., Pavithra, M., & Vishali, P. (2022). Key Object Classification for Action Recognition in Tennis Using Cognitive Mask RCNN. In Lecture Notes in Networks and Systems (Vol. 287, pp. 121–128). Springer Science and Business Media Deutschland GmbH. doi: https://doi.org/10.1007/978-981-16-5348-3_9
19. Gall, J., Yao, A., Razavi, N., Van Gool, L., & Lempitsky, V. (2011). Hough forests for object detection, tracking, and action recognition. IEEE Transactions on Pattern Analysis and Machine Intelligence, 33(11), 2188–2202. doi: https://doi.org/10.1109/TPAMI.2011.70
20. Liu, X., You, T., Ma, X., & Kuang, H. (2018). An optimization model for human activity recognition inspired by information on human-object interaction. In Proceedings - 10th International Conference on Measuring Technology and Mechatronics Automation, ICMTMA 2018 (Vol. 2018-January, pp. 519–523). Institute of Electrical and Electronics Engineers Inc. doi: https://doi.org/10.1109/ICMTMA.2018.00131
21. Yadav, P., Salwala, D., Das, D. P., & Curry, E. (2020). Knowledge graph driven approach to represent video streams for spatiotemporal event pattern matching in complex event processing. International Journal of Semantic Computing, 14(3), 423–455. doi: https://doi.org/10.1142/S1793351X20500051
22. M. H. Rahman and N. Bouguila, "Efficient Feature Mapping in Classifying Proportional Data," in IEEE Access, vol. 9, pp. 3712-3724, 2021, doi: https://doi.org/10.1109/ACCESS.2020.3047536.
23. Liz Oz, Always AI (2022), 17 interesting applications of Object Detection for businesses https://alwaysai.co/blog/object-detection-for-businesses June 4, 12.30PM PST
24. Abdellah Chehri, Hussein T. Mouftah, Autonomous vehicles in the sustainable cities, the beginning of a green adventure, Sustainable Cities and Society, Vol 51, 2019, 101751, ISSN 2210-6707, doi: https://doi.org/10.1016/j.scs.2019.101751.
25. Torrens PM. Smart and Sentient Retail High Streets. Smart Cities. 2022; 5(4):1670-1720. doi: https://doi.org/10.3390/smartcities5040085.
26. Lee J, Ahn B. Real-Time Human Action Recognition with a Low-Cost RGB Camera and Mobile Robot Platform. Sensors. 2020; 20(10):2886. doi: https://doi.org/10.3390/s20102886

Part II
Applications of Spatiotemporal Data Analytics

Chapter 7
Spatio-temporal Data Analytics for e-Waste Management System Using Hybrid Deep Belief Networks

K. Suresh Kumar, C. Helen Sulochana, D. Jessintha, T. Ananth Kumar, Mehdi Gheisari, and Christo Ananth

Abstract In the most recent few decades, there has been a significant increase all over the world in the amount of waste electronic equipment. This is a result of a number of factors, including an increase in production, management rules that are ineffective, recycling practices that are inefficient, and safety risks that are unacceptable. Because of the harmful emissions that are released into the air, water, and soil when electronic waste is disposed of in landfills, this practice has the potential to be detrimental to both human health and the environment. The process of recycling used electronic equipment results in the production of several distinct categories of secondary waste, including solids, liquids, and gases. When waste, such as electronic waste, is thrown away in an improper manner, it has a detrimental effect not only on human health but also on the environment. In this work, we propose a hybrid deep learning model (HDLM) which comprises Fuzzy-based Spatio-Temporal Optimization Mechanism (FSTOM) in addition to Deep Belief networks

K. S. Kumar
Sri Krishna college of Technology, Coimbatore, India
e-mail: sureshkumar.k@skct.edu.in

C. H. Sulochana
St. Xavier's Catholic College of Engineering, Nagercoil, India
e-mail: helentcr@gmail.com

D. Jessintha
Easwari Engineering College, Chennai, India
e-mail: jessintha.d@eec.srmrmp.edu.in

T. A. Kumar (✉)
IFET College of Engineering, Villupuram, India
e-mail: tananthkumar@ifet.ac.in

M. Gheisari
Department of Computer Science, Islamic Azad University, Shiraz, Iran
e-mail: mehdi.gheisari61@gmail.com

C. Ananth
Samarkand State University, Samarkand, Uzbekistan
e-mail: dr.christoananth@gmail.com

J. A et al. (eds.), *Spatiotemporal Data Analytics and Modeling*, Big Data Management, https://doi.org/10.1007/978-981-99-9651-3_7

(DBN) for e-waste prediction. During network training and error checking, residual learning is designed to prevent oscillations. In this chapter, we propose using a novel algorithm called MSOK (modified self-organizing map and K-Means algorithm) to create a profile for each type of waste produced. To take advantage of the best features of both statistical modelling and deep belief network modelling, these hybrid models combine the two. The findings suggest that cutting-edge data analysis techniques could be employed to obtain more accurate statistics regarding garbage generation. Compared with the existing models, the proposed model performs well effectively in determination and classification of the e-wastes with proper optimization strategies.

Keywords Deep belief networks (DBNs) · Modified self-organizing map · Health and environmental risks · Spatio-temporal optimization · e-Waste management

7.1 Introduction

Spatio-temporal data analytics is an approach that integrates the spatial and temporal dimensions of data to extract meaningful insights and patterns from complex data sets [1]. In the context of e-waste management, spatio-temporal data analytics can help identify patterns and trends in the generation, collection, and disposal of electronic waste. Hybrid deep belief networks (HDBNs) are a type of artificial neural network that combines the strengths of deep neural networks (DNNs) and Bayesian networks [2]. HDBNs are particularly well-suited for handling complex and noisy data sets, making them a powerful tool for spatio-temporal data analytics. In the context of e-waste management, HDBNs can be used to analyse large-scale data sets that capture the spatio-temporal patterns of e-waste generation, collection, and disposal. By analysing these patterns, e-waste management systems can be optimized to reduce the environmental impact of electronic waste and improve the efficiency of the waste management process [3]. Numerous spatio-temporal data sets are gathered on a daily basis across various domains. The aforementioned information can be obtained in various formats. The act of scrutinizing and extracting information from this type of data holds paramount importance in propelling the forefront of various scientific predicaments and pragmatic implementations [4]. However, it is worth noting that data possessing spatial and temporal attributes exhibit distinct properties when compared to relational sources that have been studied in conventional data mining literature. The aforementioned data exhibit temporal and/or spatial interdependence, whereby individual instances are not mutually exclusive and are not uniformly distributed. The statement suggests that there exists a possibility of structural correlation among samples in specific spatial regions or temporal instances [5]. Furthermore, it should be noted that these entities are dynamic in nature, allowing for the modification of their class attribute by individual instances at any given time and in any given location. Consequently, conventional techniques employed in data mining are not optimal for handling spatio-temporal data, resulting in inferior performance and potentially inaccurate interpretations.

The field of spatio-temporal data mining, commonly referred to as STDM, presents novel methodologies for examining event-driven data, trajectories, spatial maps, time series, and raster data. It is a relatively recent area of study [6]. STDM is commonly associated with a range of fundamental tasks. The introduction of Space-Time Division Multiplexing (STDM) has presented novel challenges and prospects in diverse fields of application. Numerous academic publications have portrayed and examined spatio-temporal data utilizing network analysis. The network representation holds significant value due to its ability to depict sub-manifolds in multi-dimensional space, capture dynamic and topological structures (such as hierarchical structures or communities), and recognize global or local patterns irrespective of data distribution [7]. The popularity of network representation can be attributed to the justification provided by its advantages. Data collected from events frequently exhibit a greater prevalence of values that are either zero or indicate the absence of an event throughout the duration of data collection. The aforementioned factor evidently affects the outcomes of the correlation analysis. The utilization of the visibility property in time series is an exemplification of an alternative approach that has been suggested for implementation in the development of networks. Notwithstanding, this methodology solely produces individual and unalterable depictions of the temporal sequence, and it neglects the geographical context [8]. An alternative approach to network construction considers the temporal sequence of spatio-temporal occurrences as a basis for establishing inter-node relationships. The researchers utilized the aforementioned tool to conduct an analysis of seismic activity. This involved constructing a network within a segmented geographical area comprised of multiple small cubic cells. The network was constructed based on the co-occurrence of events within a grid division of the area [9]. This approach was employed to analyse the fires that transpired in the Amazon basin. The aforementioned task was executed utilizing an identical approach. The authors proffered two distinct methodologies for performing data analysis, namely a temporal network and a single-layer network. Previous studies have identified valuable information in multiple domains. However, the techniques and findings presented are restricted to the specific issue under consideration and have not been extrapolated to various other domains that involve spatio-temporal events. This is the case notwithstanding the antecedent works. This research presents a holistic methodology for representing spatial and temporal events using networks as a structure for tackling challenges in data mining, such as clustering, predictive analysis, pattern recognition, anomaly identification, change detection, and relationship extraction [10]. The aforementioned approach is applicable for the depiction of spatio-temporal occurrences utilizing diverse network types. The proposed strategy involves the division of the tridimensionality space into discrete grid cells, with each of them being denoted by a corresponding node. The establishment of links between nodes is contingent upon the occurrence of a sequence of events within the corresponding cells [11]. As such, nodes are interconnected in a chronological fashion. Consequently, the hyperlinks denote the sequence of incidents in a temporal arrangement, which we shall term as a "chronnet" for the purpose of convenience. Thus, the present study extends the existing literature by proposing a generalized approach to the mechanism of connections, which leads to the conversion of spatio-temporal data into a network. Networks of this nature have been examined in other domains, including text mining and social media. Thus, the recurring sequential occurrences are depicted through robust connections that possess a significant degree of weightage. The network that was constructed exhibits noteworthy features that conform to the original spatio-temporal event distribution, including scale-free and small-world properties for power-law event distribution [12]. The aforementioned attributes are observable due to the fact that the initial distribution of events adhered to a power law. The utilization of chronnets facilitates the identification of significant nodes and anomalies, while also facilitating the exploration of an unexplored category of spatio-temporal community. Of significant note is the computational efficiency of the method, which exhibits a linear complexity and is amenable to parallelization and distribution [13]. Adjusting sequence connections, grid partitioning, and temporal windows for analyses can customize the approach to specific fields. Simulations and real-world data have been used to evaluate and analyse network construction. A simple data set generator was proposed to explain how the chronnet topology has unique spatial and temporal characteristics. This study uses a recommended data set generator to demonstrate the efficacy of chronnet measures in extracting spatio-temporal data information, as proposed in reference [14]. The data provided may encompass details such as temporal and spatial references. In order to facilitate a case study, the approach we used was implemented on a genuine dataset consisting of global fire incidents. The results suggest that our model effectively characterizes the incidence rate of fire events, areas exhibiting anomalous activity, and the seasonal patterns of fires by utilizing a solitary chronnet in tandem with intricate network metrics.

Overall, spatio-temporal data analytics using HDBNs has the potential to revolutionize the way e-waste is managed, providing a powerful tool for decision-making and policy development in this critical area of environmental sustainability.

7.2 Background and Related Work of e-Waste Management

Spatio-temporal data analytics has been used in a variety of fields, including environmental management, transportation planning, and public health. In recent years, researchers have also explored the application of spatio-temporal data analytics for e-waste management, with a particular focus on using hybrid deep belief networks (HDBNs) for this purpose. The author used spatio-temporal data analytics and HDBNs to model the spatial distribution and temporal patterns of e-waste generation, collection, and disposal in China. The study found that the use of HDBNs improved the accuracy of e-waste prediction models, allowing for better planning and management of e-waste in the country [15]. The author used spatio-temporal data analytics and HDBNs to identify the factors influencing e-waste generation and disposal in the United States. The study found that factors such as population density, income, and education level were significant predictors of e-waste generation, while policies related to e-waste management were important predictors of e-waste disposal [16]. The author used spatio-temporal data analytics and HDBNs to analyse the distribution and characteristics of e-waste in China. The study found that the use of HDBNs allowed for more accurate predictions of e-waste generation, and identified key factors such as population density, urbanization level, and industrial structure as drivers of e-waste generation and disposal.

7.2.1 Waste Monitoring-Based Data Analytics

Fundamentally, data analytics entails scrutinizing unprocessed data to identify hitherto undiscovered patterns, associations, and other valuable revelations [17]. Utilities and businesses engaged in waste management can employ analytical techniques to characterize, predict, and optimize their operations by leveraging data from monitoring systems and other available sources of information. Data analytics can be employed to examine and predict the production of waste over a specified duration, within a specific geographic region, or among a particular group of waste generators. The acquisition of data pertaining to the production of waste can establish the basis for strategic and operational decision-making in waste management practices, including the logistics of waste collection and transportation. Furthermore, analytics can provide the basis for waste producer analyses, including the identification of anomalous waste generation patterns and the profiling and categorization of waste producers. One potential application of this approach is to facilitate the identification of entities that generate waste.

At a theoretical level, the methodology of data analytics can be explicated by employing the process of knowledge discovery in databases as an illustrative example [18]. Data mining plays a significant role in this particular process. This is a compendium of data analysis techniques that possess versatility in their application to diverse objectives, and can be employed for the analysis of any given dataset. The provision of both descriptive and predictive modelling yields the highest degree of value. Descriptive modelling aims to explore the intrinsic characteristics of the data, typically accomplished through techniques such as density estimation and cluster analysis. Conversely, the aim of predictive modelling is to develop a model that can predict an output variable by utilizing a group of input variables.

The term "advanced analytics" is commonly employed to denote emerging domains that are linked with the practice of data mining. An instance of this phenomenon can be observed in the utilization of machine learning techniques for the purpose of acquiring knowledge and forecasting intricate patterns of conduct [19]. Various algorithmic techniques are employed in the field of machine learning, including support vector machines and artificial neural networks. Empirical evidence suggests that support vector machine (SVM) models exhibit superior performance and robustness compared to conventional artificial neural network (ANN) models. The aforementioned benefit can be ascribed to the observation that support vector machine (SVM) models exhibit high levels of predictive precision. The subfield of deep learning has recently surfaced within the broader domain of machine learning. This particular subdiscipline of research endeavours to construct mathematical frameworks aimed at acquiring intricate representations from vast quantities of data. These architectural designs are noteworthy in the context of managing substantial volumes of intricate data.

In comparison to conventional statistical techniques, machine learning algorithms possess the capacity to identify intricate and unfamiliar functional connections from vast quantities of data without necessitating any antecedent knowledge regarding the phenomenon under investigation. In contrast to mechanistically based modelling, this confers an advantage; nonetheless, it may yield excessively intricate black-box models with restricted mechanistic interpretation. The acceptance of data-based models in practical applications may be restricted due to a deficiency in mechanistic comprehension. Therefore, the development of data-based models for practical use should prioritize simplicity and interpretability as crucial criteria.

7.2.2 Processing Methods of Waste Monitoring Data

This study examines waste monitoring data as a series of measurements documenting the amount of waste collected at different temporal and spatial locations globally [20]. Furthermore, the dataset may comprise supplementary data linked to a specific point of measurement. During the initial stages of the procedure, it is necessary to transform the unprocessed monitoring data into a structured format that is amenable to analysis using data analytics techniques. The quantification of waste production can be represented through a continuous time series, contingent upon the sampling resolution, rather than relying on discrete timestamps to denote the amount of waste generated. By integrating data obtained from weighing with information from other sources, it becomes feasible to generate indicators that exhibit a higher level of intricacy. An instance of waste generation can be articulated in the context of extrinsic factors.

Monitoring data can also approximate waste management metrics. Integration of waste volume, composition, and transportation distance data with TCS information yields site-specific transportation metrics. An efficiency index could identify waste management inefficiencies. This index uses waste collection and route data. Use waste mass, transportation, and other processes. Several methods can achieve this goal. For regional biogas production spatial mass and energy balance calculations, this approach is required. The proposed method can estimate site-specific waste management costs and effects. Thus, it can help identify the most viable and environmentally friendly solution. This study shows that segregating biowaste for collection may not be sustainable economically or environmentally in some areas. To understand waste management's full effects, life-cycle assessment is needed. LCA does not prioritize local impacts, but waste management decisions should.

7.2.3 Construction of Predictive Models Dedicated for Waste Generation

The production of waste is a multifaceted outcome that arises from a convoluted interplay among consumer behaviour, purchased commodities, the natural surroundings, and diverse other determinants, which is often misconstrued. The non-linear, discontinuous, and spatio-temporal nature of certain relationships can render conventional statistical methods inadequate for analysis and modelling purposes.

The precision of temporal measurements is contingent upon the employed measurement methodology and the rate of sampling. In instances where on-board monitoring is not conducted continuously, the temporal resolution is contingent upon the duration between waste collections. Consequently, the data obtained can solely be utilized to detect significant seasonal variations and rhythms. Real-time sensor measurements are necessary to capture hourly or daily variations in waste accumulation. The waste generation data are the monitoring data utilized in the compilation of a time series. The aforementioned data can be subjected to analysis, modelling, and prediction through the application of methodologies such as regression and time-series analysis.

Predictive models can be developed by utilizing waste quantity data that is being monitored along with relevant external data. Scholarly literature encompasses a diverse range of alternative approaches, including conventional regression-based methods and more sophisticated machine learning techniques. A total of 45 distinct modelling methodologies have been identified for the purpose of elucidating or predicting the current or future production of municipal solid waste across various scales, including the household, district, settlement area, and country levels. These approaches rely on the utilization of economic, socio-demographic, or management-related data. In contemporary practice, machine learning techniques are progressively employed alongside conventional regression methodologies. These methodologies endeavour to ascertain a non-linear function that exists between waste generation and the explanatory factors. The extant literature contains numerous case studies that illustrate the efficacy of artificial neural network (ANN) models in predicting the quantity of waste generated.

7.3 Waste Generation Profiling

In the context of descriptive waste analytics, the utilization of a clustering algorithm makes it possible to categorize waste sites and waste producers into groups that display waste generation patterns that are comparable to one another. It is possible to develop a series of waste generation profiles that accurately reflect the typical behaviour of waste generation within a particular group if one first classifies the patterns of waste generation that occur within a particular group. As a consequence of this, it is clear that every location that generates waste belongs to a particular category, which can be recognized on the basis of the type profile that it possesses. It is possible to conduct an analysis and determine the underlying factors that contribute to waste generation within each group if waste producer-related data, such as household type, income, and education level, are used. In addition to this, the information can be put to use in order to classify newly produced waste into the previously established categories [21].

Innovative information services, such as comparative analyses of waste generation between individual households and waste producers, are something that can be created by drawing on the insights that are provided by descriptive analytics. These insights can be used. The use of comparative data can be helpful in facilitating the identification of atypical patterns of waste generation, which in turn enables the development of customized strategies for the management of waste. These strategies include hierarchical and conventional clustering techniques, such as K-Means and ISODATA, respectively. Hierarchical clustering is an example of a methodology that was used, and it was used in order to identify relevant socioeconomic characteristics of municipalities in relation to the production of household waste. In order to examine the amount of mixed waste produced by households across all of the different types of housing, descriptive statistics and distribution were utilized. In the field of electricity distribution, data-driven methods are becoming increasingly common for constructing electric load profiles that are individualized to individual building sites through the use of monitoring data. The proposed method makes use of the ISODATA methodology to generate a collection of index series that correspond to various types of days. These index series are then normalized with regard to annual energy and outdoor temperature. Utilize an algorithmic approach based on K-Means and self-organizing maps (SOM) to build the corresponding index series. In general, the findings of these investigations highlight the potential of spatio-temporal data analytics as well as HDBNs to improve existing e-waste management systems. By providing more accurate forecasts of e-waste production and identifying significant factors that influence e-waste disposal, the utilization of these methodologies can assist in the development of policies and decision-making processes concerning environmental sustainability. This can be accomplished through the identification of significant factors that influence e-waste disposal.

7.3.1 Hybrid Deep Learning Model for e-Waste Prediction

E-waste prediction is an important task that can help in managing electronic waste effectively. A hybrid deep learning model can be used to predict e-waste generation by combining different deep learning architectures. Here's a possible approach.

Data collection: Collect data on e-waste generation from various sources such as waste management companies, electronic manufacturers, and government reports.

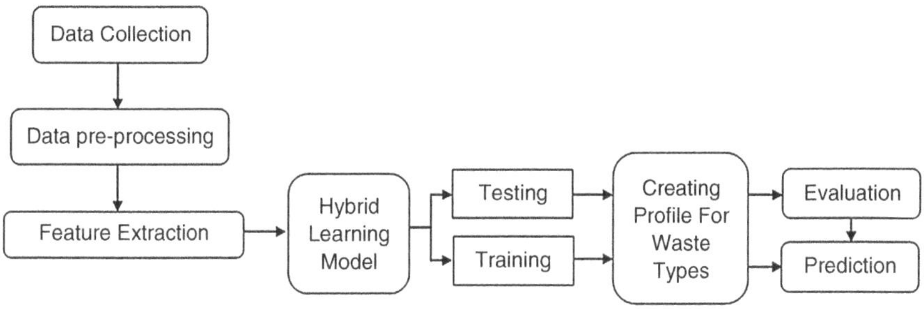

Fig. 7.1 Architecture of the proposed model

The data should include information on the type of electronic devices, their lifespan, and the rate of disposal. The architecture of the proposed model is shown in Fig. 7.1.

Data pre-processing: Clean the collected data and perform necessary transformations such as normalization and encoding.

Feature extraction: Extract relevant features from the pre-processed data such as device type, lifespan, and rate of disposal.

Hybrid deep learning model: Build a hybrid deep learning model that combines different deep learning architectures such as Fuzzy-based Spatio-Temporal Optimization Mechanism in addition with deep belief networks. The hybrid model can use the extracted features as inputs to the different architectures.

Training: Train the hybrid deep learning model using the pre-processed data. Use appropriate techniques such as dropout and regularization to prevent overfitting.

Creating profile for waste types: A novel algorithm is proposed, namely MSOK, to create a profile for each type of waste produced.

Evaluation: Evaluate the trained model on a test dataset to determine its accuracy and performance.

Prediction: Use the trained model to predict e-waste generation based on the input features.

By combining different deep learning architectures in a hybrid model, we can take advantage of their respective strengths and achieve better performance in e-waste prediction. However, it's important to note that the success of the model depends on the quality and quantity of the data used for training.

7.3.1.1 Fuzzy-Based Spatio-Temporal Optimization Mechanism

A Fuzzy-based Spatio-Temporal Optimization Mechanism is a computational approach that uses fuzzy logic and spatial-temporal optimization techniques to solve problems related to space and time. Here's an overview of how this mechanism works:

Problem formulation: The first step is to formulate the problem as an optimization problem with spatio-temporal constraints. For example, the problem could be to find the optimal route for a vehicle that takes into account traffic conditions and travel time.

Spatial-Temporal Optimization: Spatial-Temporal Optimization techniques are used to solve the optimization problem. These techniques involve finding the optimal solution while taking into account the spatial and temporal constraints.

Fuzzy-based optimization: Fuzzy logic is used to modify the optimization process to take into account uncertain and imprecise data. For example, fuzzy logic can be used to adjust the weights assigned to different constraints based on their importance.

Evaluation: The results of the optimization process are evaluated to determine their quality and effectiveness.

The utilization of a Fuzzy-based Spatio-Temporal Optimization Mechanism is a potent approach for addressing issues pertaining to both spatial and temporal domains. This technology exhibits versatility in its applicability, spanning various domains including transportation planning, logistics, and environmental management. The utilization of fuzzy logic for modelling uncertainty and imprecision, coupled with spatial-temporal optimization techniques for identifying the optimal solution, can effectively address intricate problems that may prove challenging or unfeasible to resolve through conventional means.

A Fuzzy-based Spatio-Temporal Optimization Mechanism can be used in e-waste management to optimize the logistics and transportation of electronic waste by considering the spatial and temporal aspects of the problem. This mechanism can be used to identify the most efficient and cost-effective routes for collecting and transporting electronic waste, taking into account factors such as traffic patterns, waste generation rates, and the capacity of the recycling facilities.

The Fuzzy-based Spatio-Temporal Optimization Mechanism can incorporate fuzzy logic to handle uncertain and imprecise data, such as traffic conditions or waste generation rates. By considering these uncertain factors, the mechanism can generate more accurate and reliable optimization solutions. The mechanism can also use machine learning techniques, such as deep learning, to learn from historical data to improve the accuracy of predictions and optimize future routes. The mechanism can optimize the collection and transportation of electronic waste in real time by continuously analysing and processing data from various sources. For example, it can use data from GPS trackers on collection vehicles, traffic monitoring systems, and waste generation sensors to update the optimization model and generate new solutions as the situation changes. Moreover, the mechanism can be designed to minimize the environmental impact of the transportation process by considering factors such as the carbon footprint of the vehicles used and the distance travelled. The Fuzzy-based Spatio-Temporal Optimization Mechanism can optimize the routes based on environmental criteria and generate the most environmentally friendly solutions for electronic waste transportation. Fuzzy-based Spatio-Temporal Optimization Mechanism can be a valuable tool for e-waste management, helping to optimize the logistics and transportation of electronic waste by considering spatial and temporal aspects of the problem, incorporating fuzzy logic to handle uncertainty and imprecision, and minimizing the environmental impact of the transportation process.

7.3.2 Formation of Dynamic Spatio-temporal Network

The research utilized the periodic shifting attention mechanism and the flow gating mechanism to gain an understanding of the dynamic spatio-temporal dependencies. The predictive capacity for the two distinct traffic volumes was enhanced through the incorporation of the start and end volumes. The aforementioned measure was implemented based on previous studies that have demonstrated remarkable effectiveness in the field of traffic prediction. The Fuzzy-based Spatio-Temporal Optimization mechanism has been identified as a constituent element of our proposed model. Previous research has utilized convolutional neural networks (CNNs) to investigate the spatial interdependencies between separate regions. The urban environment, when viewed as a grid, reveals that the utilization of CNN as the sole analytical tool is inadequate in comprehensively capturing the intricate interrelationships between the central area and its surrounding regions. The utilization of the flow gating mechanism endures as a method for acquiring understanding regarding the dynamic spatial interrelationships within the hierarchical structure. It is commonly recognized that each geographic area maintains information regarding the start and end of the complete dataset. The target region is marked to be i and also the surrounding regions are marked as K×K images along with two channels represented mathematically as

$$\mathrm{T}_{t,i} \in Y^{K\times K\times 2} \text{ at each time interval "} t \text{".} \tag{7.1}$$

Using data collected over a period of s discrete time intervals, a stacked depiction of the regional flow matrix can be constructed.

$$\Psi_{t,i} \in Y^{K\times K\times 2s} \tag{7.2}$$

The variable "s" denotes the quantity of flow matrices. The utilization of convolutional neural networks (CNNs) enables the concurrent modelling of both spatial interaction and spatial flow interaction among regions. Subsequently, the flow information can be utilized to implement the flow gating mechanism, which has the capability to apprehend the fluctuating interdependencies among the areas via the compression of spatial feature information. The formula of the output $\mathrm{T}_{t,i}$, $\Psi_{t,i}$ of the spatial representation modulated by flow gating for each layer is as follows:

$$T^{(m)}{}_{t,i} = \sum_{x\in R\to\alpha} R_{i,t}(x) + \left(\mathrm{Re}LU(N_f{}^{(m)} * T_{i,t}{}^{(x-1)} + \mathrm{P}_f{}^{(m)}\right) \tag{7.3}$$

$$\Psi^{(m)}{}_{t,i} = \sum_{x\in R\to\alpha} R_{i,t}(y) + \left(\mathrm{Re}LU(N^{(m)} * \Psi_{i,t}{}^{(y-1)} + \mathrm{P}^{(m)}\right) \tag{7.4}$$

The parameters utilized in this context are commonly referred to as the acquired parameters. In contrast to the approach presented in this chapter, which employs an equal number of gated and local convolutional layers, the proposed method separates gated convolution operations from local convolution operations. This structural enrichment of the model may lead to improved performance. Additionally, it has the potential to decrease the computational time required due to the intricacy of the model in specific situations. Subsequently, a flat layer and a fully connected layer are sequentially incorporated to obtain the spatial feature representation of the flow gated as $y_{i,t}$.

7.3.3 Deep Belief Networks

Deep belief networks (DBNs) belong to a category of deep neural networks that are predominantly utilized for unsupervised learning purposes, including but not limited to dimensionality reduction, feature learning, and generative modelling. Deep belief networks (DBNs) are constructed by arranging numerous layers of restricted Boltzmann machines (RBMs) in a hierarchical manner.

Deep belief networks are a type of deep learning algorithm that can be used for a wide range of applications, including e-waste management. E-waste management refers to the proper disposal and recycling of electronic waste, which is becoming an increasingly important issue due to the rapidly growing use of electronic devices. One way that DBNs can be used for e-waste management is by helping to identify and classify different types of electronic waste. For example, a DBN could be trained on a large dataset of images of electronic waste, such as discarded phones, computers, and other devices, and then used to automatically classify new images of e-waste. This could help make the process of sorting and recycling e-waste more efficient and cost-effective.

DBNs could also be used to predict which electronic devices are more likely to be discarded or replaced in the near future. By analysing data on consumer behaviour and device usage patterns, a DBN could identify patterns that indicate when certain devices are likely to be disposed of or replaced. This could help manufacturers and retailers better understand the lifecycle of their products and develop strategies to reduce waste and promote sustainability. In addition to these applications, DBNs could also be used to optimize the process of dismantling and recycling electronic waste. By analysing the structure and components of different electronic devices, a DBN could identify the most efficient and effective ways to disassemble and recycle them. Overall, DBNs have the potential to be a valuable tool for e-waste management, helping to reduce waste, promote sustainability, and improve the efficiency of the electronics industry.

7.3.4 Fuzzy-Based Deep Belief Networks

Fuzzy-based deep belief networks are a type of deep neural network that combines fuzzy logic and deep learning techniques. FDBNs can be used for a variety of applications, including classification, regression, and clustering. Fuzzy-based deep belief networks can be used for e-waste management by combining the strengths of fuzzy logic and deep learning. Fuzzy logic is a type of logic that allows for imprecise or uncertain inputs, which is useful in the context of e-waste management where there may be variations in the type and condition of electronic devices being recycled.

One way that F-DBNs can be used for e-waste management is by improving the accuracy of classification and identification of electronic waste. Fuzzy logic can help to deal with the uncertainty in the data, and the DBN can be used to learn complex representations of the data that can improve the accuracy of classification. By combining the two techniques, F-DBNs can improve the accuracy of identifying different types of electronic waste, which can help to streamline the recycling process. F-DBNs can also be used to optimize the dismantling process of electronic waste by considering uncertain factors such as the presence of hazardous substances. Fuzzy logic can help to handle imprecise data about the composition of electronic devices, which can then be used by the DBN to generate recommendations for dismantling and recycling processes. Another application of F-DBNs in e-waste management is in predicting the potential of reuse of electronic devices. By analysing data on the history of device usage, maintenance, and repair, an F-DBN can predict the likelihood of reuse of an electronic device. This can help to promote the reuse of devices and reduce the amount of electronic waste generated.

7.3.5 MSOK

Modified Self-Organizing Map and K-Means algorithm are two popular clustering techniques that have been modified to improve their performance. Unsupervised neural network algorithm Self-Organizing Map (SOM) projects high-dimensional data onto a low-dimensional space. Analogous data points in a reduced-dimensional space achieve the goal. The popular K-Means algorithm clusters data points into K clusters based on their proximity to K centroids. Here's an overview of how the modified SOM and K-Means algorithms work:

7.3.5.1 Modified Self-Organizing Map

Initialization: The modified SOM algorithm initializes the weights of the SOM nodes randomly.

Neighbourhood Function: The neighbourhood function is used to update the weights of the SOM nodes based on the distance between the input data point and the nodes. The modified SOM algorithm uses a Gaussian function for the neighbourhood function.

Learning Rate: The learning rate is used to control the rate at which the SOM nodes are updated. The modified SOM algorithm uses a variable learning rate that decreases over time.

Clustering: After the weights of the SOM nodes have been updated, the data points are clustered based on their distance from the SOM nodes.

A modified self-organizing map can be used for e-waste management by creating a mapping of the different types of electronic waste to facilitate the sorting and recycling process. SOM is a type of artificial neural network that can identify patterns and similarities in large datasets. In e-waste management, SOM can be trained on a large dataset of images of electronic waste to create a visual map of the different types of electronic waste. The modified SOM can incorporate additional features, such as the condition of the electronic devices or the presence of hazardous materials, to create a more detailed and accurate mapping. The modified SOM can be used to facilitate the sorting and recycling process by identifying the type of electronic waste and providing information on the most appropriate recycling process. The map generated by the modified SOM can be used as a reference guide for workers in e-waste recycling facilities to quickly and accurately identify the type of electronic waste and determine the most appropriate recycling process. In addition to facilitating the sorting and recycling process, the modified SOM can also be used to identify areas with high concentrations of electronic waste. By analysing data on the location and distribution of electronic waste, the modified SOM can identify hotspots of electronic waste accumulation, which can then be used to inform targeted recycling programs.

Moreover, the modified SOM can be continuously updated with new data to improve the accuracy of the mapping and adapt to changes in the types and conditions of electronic waste. This can help to improve the efficiency of the sorting and recycling process and reduce the environmental impact of e-waste disposal.

7.3.5.2 Modified K-Means Algorithm

The utilization of a modified K-Means algorithm presents a viable approach for enhancing the clustering and sorting of electronic waste in the context of e-waste management. The K-Means algorithm is a clustering technique that partitions a given dataset into k clusters according to their similarity. The modified K-Means algorithm has been identified as a viable approach in the realm of e-waste management, whereby electronic waste can be effectively grouped based on shared characteristics such as material composition, size, and type.

One modification to the K-Means algorithm for e-waste management is to include additional features that are specific to electronic waste, such as the presence of hazardous materials, condition of the electronic device, and the ease of disassembly. By incorporating these additional features into the algorithm, the modified K-Means algorithm can improve the accuracy of clustering and sorting of electronic waste, which can streamline the recycling process and reduce the environmental impact of e-waste disposal.

Another modification to the K-Means algorithm is to use a hierarchical clustering approach, where clusters are created recursively by dividing the data into smaller groups. This can be useful for e-waste management because it can handle the complexity and diversity of electronic waste, which may require more nuanced and fine-grained sorting than a traditional K-Means algorithm.

Additionally, the modified K-Means algorithm can be used to optimize the dismantling process of electronic waste by grouping electronic devices into clusters that are easier to disassemble. By considering the ease of disassembly as a feature, the modified K-Means algorithm can generate clusters that are optimized for efficient and safe dismantling, which can reduce the time and cost of the recycling process.

Moreover, the modified K-Means algorithm can be continuously updated with new data to adapt to changes in the types and conditions of electronic waste. This can help to improve the accuracy of clustering and sorting and ensure that the recycling process is efficient and effective. The overall process involved in the e-waste management using K-Means algorithm is discussed in Fig. 7.2.

Initialization: The modified K-Means algorithm initializes the centroids of the K clusters randomly.

Neighbourhood Function: The neighbourhood function is used to update the centroids of the K clusters based on the distance between the input data point and the centroids. The modified K-Means algorithm uses a Gaussian function for the neighbourhood function.

Learning Rate: The learning rate is used to control the rate at which the centroids of the K clusters are updated. The modified K-Means algorithm uses a variable learning rate that decreases over time.

Clustering: After the centroids of the K clusters have been updated, the data points are clustered based on their distance from the centroids.

The modified SOM and K-Means algorithms have been shown to improve the performance of the original algorithms in terms of clustering accuracy and convergence speed. These modifications are particularly useful for high-dimensional data where traditional clustering algorithms can struggle to find meaningful clusters.

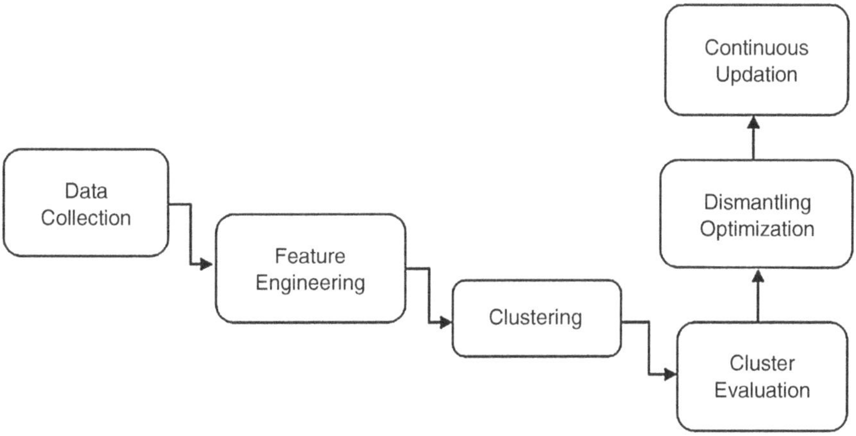

Fig. 7.2 e-waste management using K-Means algorithm

7.4 Analysis of e-Waste Management

7.4.1 Dataset Description

In e-waste management systems using deep learning, the dataset typically consists of various types of electronic waste and associated data, such as material composition, size, type, condition, and potential hazards. The dataset may also include images or other sensory data to provide a more comprehensive understanding of the electronic waste. In this approach, some of the datasets used for the training purposes are taken from the open database Kaggle.com. In a spatio-temporal data analytics system for e-waste management, the dataset typically includes data on the spatial and temporal distribution of electronic waste. These data can be obtained from various sources, such as e-waste recycling facilities, manufacturers, and government agencies. The dataset may include the following features:

Geographic Information: The dataset may include spatial data such as the location of e-waste collection points, recycling facilities, and landfills. These data can be represented in the form of geographic coordinates or maps.

Time Series Data: The dataset may include time series data that captures the temporal distribution of electronic waste over a period of time. These data can be represented in the form of time stamps or dates.

Material Composition: The dataset may include information on the material composition of the electronic waste, such as plastics, metals, and other components.

Type: The dataset may include information on the type of electronic devices, such as smartphones, laptops, and televisions.

Condition: The dataset may include information on the condition of the electronic devices, such as whether they are functional, damaged, or outdated.

Hazardous Materials: Toxins such as lead, mercury, and cadmium are just a few examples of the kinds of substances that might be contained in the collection. These substances pose a risk to the health of humans as well as the environment.

Transportation Routes: The dataset may include information on transportation routes for electronic waste, such as the routes taken by e-waste collection trucks and the distances travelled.

Demographic Data: The dataset may include demographic data such as population density, income levels, and other socioeconomic factors that can impact the generation and disposal of electronic waste.

Environmental Data: The dataset may include environmental data such as air quality, water quality, and other environmental factors that can impact the disposal and recycling of electronic waste.

7.5 Results of Categorizing Some Types of Electronic Trash

The HDLM network's initial configuration was written in Python using the developer-supplied samples. Table 7.1 lists details about the two options that were considered.

Table 7.1 Classification results for HDLM configurations

	Deep learning CNN configuration							Accuracy	
Configuration number	Input	C1	C2	C3	C4	F1	F2	Test images	Training images
HDLM1	54	32 FM FS 7 × 7 MS 20 × 20	32 FM FS 8 × 8 MS 26 × 26	32 FM FS 6 × 6 MS 29 × 29	32 FM FS 5 × 5 MS 19 × 19	256	1	0.93	0.96
HDLM2	54	33 FM FS 9 × 9 MS 30 × 30	33 FM FS 2 × 2 MS 32 × 32	33 FM FS 31 × 31 MS 34 × 34	33 FM FS 23 × 23 MS 35 × 35	256	4	0.92	0.89
HDLM3	65	34 FM FS 10 × 10 MS 60 × 60	34 FM FS 11 × 11 MS 62 × 62	34 FM FS 12 × 12 MS 63 × 63	34 FM FS 13 × 13 MS 65 × 65	256	6	0.86	0.92
HDLM4	34	35 FM FS 11 × 11 MS 120 × 60	35 FM FS 12 × 12 MS 120 × 60	35 FM FS 15 × 15 MS 120 × 60	35 FM FS 16 × 16 MS 120 × 60	256	8	0.98	0.93
HDLM5	23	36 FM FS 12 × 12 MS 120 × 120	36 FM FS 18 × 18 MS 128 × 128	36 FM FS 20 × 20 MS 108 × 108	36 FM FS 22 × 22 MS 126 × 126	256	10	0.84	0.93
HDLM6	45	37 FM FS 20 × 20 MS 60 × 60	37 FM FS 26 × 26 MS 160 × 160	37 FM FS 27 × 27 MS 260 × 260	37 FM FS 22 × 22 MS 150 × 150	256	12	0.85	0.94
HDLM7	34	38 FM FS 24 × 24 MS 240 × 60	38 FM FS 32 × 32 MS 220 × 220	38 FM FS 36 × 36 MS 150 × 150	38 FM FS 38 × 38 MS 80 × 80	256	14	0.82	0.91
HDLM8	65	39 FM FS 36 × 36 MS 240 × 240	39 FM FS 45 × 45 MS 232 × 232	39 FM FS 62 × 62 MS 210 × 210	39 FM FS 72 × 72 MS 230 × 230	256	16	0.86	0.95

FS Filter size, *FM* Feature map, *MS* Feature map size, max pooling with × filter

7.5.1 Recognition of Several Objects in an Image

In order to accomplish the simultaneous recognition of multiple objects contained within a single image, the use of a Faster Feature Pyramid-Based Network (FDBN) was necessary. The FDBN method is a technique for the detection of objects that entails the selection of particular regions for processing and, subsequently, the determination of the classification of those regions. This method is known as the FDBN method. The FDBN methodology makes it possible to identify a wide variety of object categories and determine the respective dimensions of those object categories within the provided image. The act of specifying the dimensions of objects makes it possible to compute proportions, and as a result, the determination of the magnitude of the devices in question is made possible. Figure 7.3 illustrates the results obtained when the faster R-CNN algorithm was applied to an image that included a pair of washing machines, a refrigerator, and a monitor.

The object detection model known as Faster FDBN made use of a Convolutional Neural Network (CNN) architecture. This architecture was made up of three convolutional layers that each contained filters of a different dimension. In order to achieve the best possible performance, the configuration of the model was fine-tuned. The completion of each of the four stages of network training with a predetermined minimum number of epochs was required in order to meet the requirements for a successful learning completion. Even though it was not required that all of the images in the training dataset have the same dimensions, the FDBN was trained on a consistent set of pictures. The application of accelerated feedforward deep belief networks (FDBN) makes it easier to recognize objects and the dimensions that are

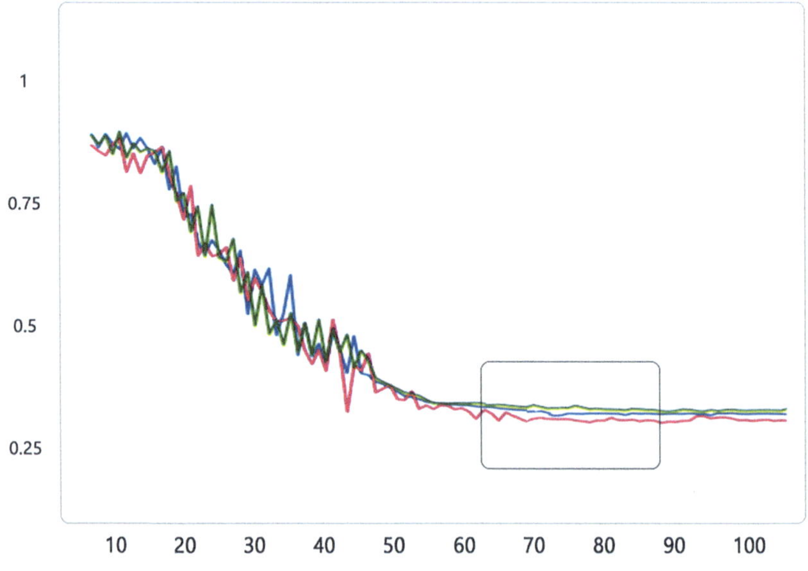

Fig. 7.3 Results obtained when the faster R-CNN algorithm

associated with them within visual representations. When it comes to the planning of waste collection, having a good understanding of the dimensions of the equipment is important. The selected machinery belongs to the group known as Waste Electrical and Electronic Equipment (WEEE), which is distinguished by a substantial weight and a diverse range of dimensions. The machinery was selected because it fits the criteria for this category. A representation of waste equipment has the potential to encompass a variety of different configurations, such as the presence of one, two, or three distinct categories of electronic waste, as well as multiple instances of the same equipment or equipment belonging to the same classification. During the first stage of the planning process for the collection of electronic waste, it is necessary to determine the dimensions of the items in addition to determining the category of the equipment. In order to put into action a strategy for the collection of waste electrical and electronic equipment (WEEE), it is necessary to set aside an appropriate vehicle that is equipped with the capability of transporting all of the discarded equipment in accordance with the requirements of the customer. Customers and collection companies both benefit from the implementation of an automated system for the reporting of e-waste because of the convenience it provides. Customers are obligated to take and send in images, while the server operated by the collection company is able to identify and classify images based on the size and nature of each image. Figure 7.4 presents further examples of visuals that contain the identified objects. These examples can be found in the previous section.

The method had an accuracy rate of approximately percent overall, with a recognition rate of approximately percent for images of washing machines and approximately percent recognition rate for images of refrigerators. When compared to the

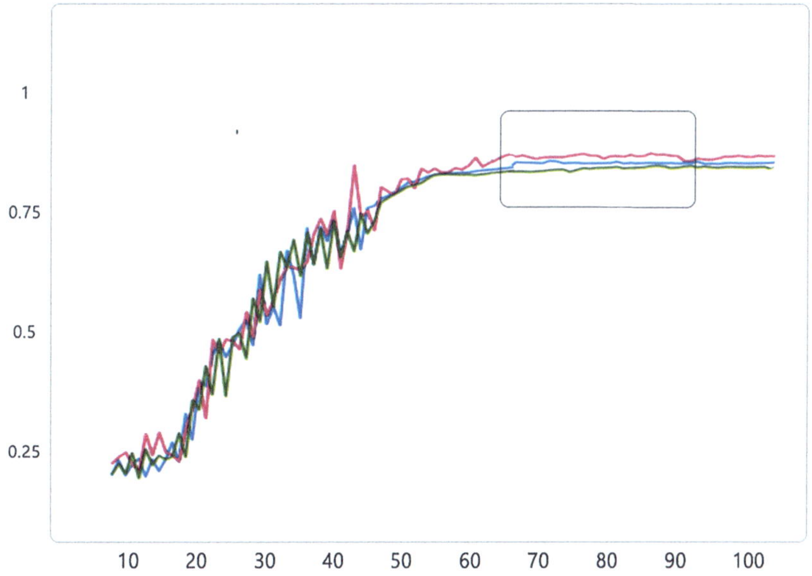

Fig. 7.4 Examples of visuals that contain the identified objects

FDBN network, the FDBN network demonstrated a significantly higher degree of accuracy in object recognition. The capability of FDBN to accurately determine the dimensions of the detected entity is one of the many advantages offered by this technology. The length of time necessary to train a neural network is determined by a number of factors, including the configuration of the network, the size of the training dataset, and the computational capacity of the machine. Python software was utilized on a personal computer that featured an Intel CoreTM i CPU running at GHz and GB of RAM in order to calculate the amount of time required for the FDBN and FDBN network to complete the learning process. According to the findings, the amount of time required to learn the FDBN was approximately minutes, while the amount of time required to learn the FDBN was approximately minutes. After receiving training, the neural networks demonstrated a high level of operational efficiency and were able to be effectively applied in real-time scenarios. The current research presents a novel method that makes use of a Feedforward Deep Belief Network (FDBN) with the intention of identifying waste equipment used in residential waste collection. The concept that has been proposed involves making use of the sharing of digital images as a way to facilitate the exchange of information between a waste collection agency and a requester who is looking for waste removal services. Another advantage of utilizing this system is the widespread adoption and general acknowledgement by society of digital solutions and mobile applications. Residents would be able to request waste retrieval in a convenient manner by submitting a photograph of the waste equipment if automated visual waste recognition technology were to be implemented. This would yield advantages for both waste collection entities and residents. Both the algorithm and the application that goes along with it have the capacity to run in either a server environment or on mobile devices using the corresponding application. Previous applications that have the potential to facilitate automated and robotic waste management systems have been improved upon, and the system that has been proposed for the image recognition and classification of e-waste demonstrates this advancement. Shaukat suggested making use of robotic platforms in order to automate the treatment of nuclear waste material. The primary objective of the research was to investigate reliable and efficient methods that are capable of being used for the visual classification of waste. In addition, it was suggested that a new machine vision system be utilized for the purpose of independently identifying waste. This proposition calls for the rotation of, and scaling up to, certain values. The use of invariant moments was put to work in order to characterize object shape, and a learning algorithm based on random forest was put to use in order to carry out object classification. The end result was a mean true positive rate of that equalled.

Managers of collection companies have voiced their displeasure with the waste equipment that is recovered, citing the fact that it frequently deviates from the specifications provided by the customer as the source of their dissatisfaction. This phenomenon makes it difficult to load vehicles, particularly in situations in which the dimensions of the waste appliances turn out to be larger than expected. The utilization of standardized shipment units in the supply chains of businesses that deal in waste collection is a distinguishing feature of this industry. Examples of standardized shipment units include pallets and cardboard boxes. When it comes to the collection and loading of electronic waste, each unit is disposed of in a random manner, and as a result, the sizes of the items that are collected vary. This is because a unit's product life can come to an end due to breakage or other factors. When a customer communicates with a company that deals with waste management, it is highly unlikely that the customer is aware of the quantity of waste they produce. Their primary objective is to dispose of an unnecessary item that is currently located in their residence. Similar predicaments might occur when there are multiple entities that need to be disposed of at the same time. It is imperative for a company to engage in effective communication with residents in order to ascertain the number of items that need to be collected in order to mitigate any potential challenges that may arise during the process of collection. This information will allow the company to determine how many items need to be collected. Because of this, the company will be able to allocate sufficient resources, such as the deployment of suitable vehicles and adequate staffing. Residents are able to communicate with the waste collection company in a number of straightforward ways, one of which is through the visual identification of waste. The primary objective of the methodology that has been proposed is to categorize and estimate the dimensions of the waste equipment that is meant to be recovered. It is not necessary to achieve a high degree of precision in size recognition because all that is required is to determine the dimensional requirements of the collection vehicles. In practice, this means that achieving this level of precision is not necessary. At this time, there is no possibility of automating the waste equipment retrieval and collection process planning. As a consequence of this, the costs that are associated with mobile collection are substantial, which has a negative impact on how collection businesses evaluate this strategy. It has been discovered that the use of automated algorithms for image recognition and classification can improve the accuracy of size estimation as well as the efficiency of type categorization. In addition, the automation feature makes it possible to recover value from waste and to plan loads in a dependable manner.

7.5.2 Analysing Trash Monitoring Data to Identify Waste Generation Profiles

The profiles of waste generation kinds were determined by a cluster analysis of the collected monitoring data. This action was executed in adherence to the previously discussed principles. The present investigation employed a modified variant of the conventional K-Means algorithm in combination with a modified version of the self-organizing map algorithm to cluster the data. The aforementioned procedure facilitated the computation of waste generation type profiles that are specific to each cluster. The combination of MSOM and modified K-Means has demonstrated successful application in various domains. For instance, it has been employed to generate electric load profiles by processing large volumes of data obtained from smart metres. At its fundamental level, MSOM serves as an initial measure to decrease the quantity of data lines. In cases where there is a large amount of waste monitoring data being analysed, the computational workload can be notably decreased by utilizing a reduced number of data lines for computation. Furthermore, the utilization of MSOM facilitates the examination of the information contained within the dataset denoted as D, thereby proving advantageous in terms of interpretation.

7.5.3 *Self-organizing Map*

The MSOM is an unsupervised learning algorithm that is applicable to various data-mining tasks in principle. The fundamental concept of MSOM involves the creation of a transformation that maps the input space, which is of high dimensionality, to an output space of lower dimensionality, typically comprising an array of neurons with a dimensionality of D. The weight vectors wi of each neuron i are initialized and adjusted via an unsupervised learning process, typically utilizing the Euclidean distance ||x-wi|| to identify the optimal matching unit b for a given random data vector. The weight vectors of the Best Matching Unit (BMU) wb and its adjacent neurons are modified through the utilization of a Gaussian function.

$$\mathrm{K}_i(t+1) = K_i(t) + \lambda_{bi}(t)\sum_{i=1}^{8}\left[x_i(t) - K_i(t)\right] \tag{7.5}$$

$$\lambda_{bi}(t) = \chi(t)\sum_{i=1}^{8}\sum_{b=1}^{8}\mathrm{H}(\mu_b - \mu_i)/\sigma \tag{7.6}$$

where a stands for the learning rate, t for the size of the iteration step, *b* and *i* for the location on the map of unit *b*, *bi* for the Gaussian neighbourhood function, and *H* for the size of the kernel. In this investigation, the MSOM was implemented using a hexagonal lattice and an *x*-sized map.

7.5.4 *The Modified K-Means Algorithm*

After the MSOM map training was finished, the MSOM prototype vectors went through a cluster analysis process that involved the application of the K-Means algorithm. When performing cluster analysis, the K-Means algorithm is a method that is frequently used. Its primary goal is to divide a set of data points into k clusters in such a way that each data point is assigned to a cluster based on how close it is to the mean of the cluster. The task of determining an appropriate number of clusters is one that requires exploration and is dependent on the particular clustering technique that is used. In the context of K-Means clustering, the Davis-Bouldin index has been utilized as a metric to evaluate the efficacy of clustering in an effort to determine whether or not it is effective.

$$DavisBouldin = \frac{1}{N}\sum_{i=1}^{M}\sum_{j=1}^{N}\max_{i,j}\frac{R_i + R_j}{d_{ij}} \tag{7.7}$$

When there are N clusters, we use their centroids to determine the distances within and between them:

$$S_i = \frac{1}{|C_i|}\sum_{i=1}^{N} x_i - l_i \tag{7.8}$$

$$S_j = \frac{1}{|C_j|}\sum_{j=1}^{N} x_j - l_j \tag{7.9}$$

where "*mi*" denotes the location of the cluster's centre and "*C*" indicates the total number of points in the cluster. Due to the fact that the K-Means method is a stochastic one, several repeated runs are essential in order to achieve sufficient significance. As a result, in order to reduce the amount of uncertainty brought on by the stochasticity, we used the K-Means clustering algorithm with a range of different numbers of clusters, or k, each of which was calculated independently. The Davis-Bouldin index indicates that there are multiple potential optima that can be found. Despite the fact that the quality of the clustering typically improves with increasing numbers of clusters, it is common practice to favour the smallest possible number of clusters.

The computed DB-index suggests that the selection of clusters should be the first option to consider because the index has reached a point where it has reached a local minimum. Clusters are another alternative that could be considered if additional profiles and a more in-depth analysis of waste generation are of interest. In terms of the SSE of K-Means clustering, both of these options can be considered valid choices.

7.5.5 Using Spatial Data That Are Already Accessible to Profile Trash Generation

The waste generation profiles generated depict diverse trends in waste generation; however, they do not explicate the underlying factors contributing to these trends. Through the integration of monitoring data with other publicly available data sources, such as demographic data, it becomes possible to elucidate and contextualize the profiles. Comprehending the rationales behind the production of heterogeneous waste in the context of waste management practices is imperative. This comprehension holds significance in the formulation of customized management strategies for specific regions and producer groups, encompassing waste collection planning, waste minimization, and recycling operations. The present investigation involved an analysis of waste generation profiles utilizing socioeconomic x-m grid data obtained from the Helsinki Metropolitan area's designated target region. Mean values for key socioeconomic variables across geospatial grids were used to merge grid data and geospatial waste generation series. Next, a self-organizing map (SOM) and K-Means-based cluster analysis was performed using additional socioeconomic input data. The socioeconomic variables included age, income, size, surface area, living space, employment rate, one-family house rate, and owner-occupied flat rate. To simplify analysis and reduce correlation, additional variables were excluded. Cluster C waste production peaks in spring. Cluster C generates most in autumn. Cluster C waste production spikes in August, possibly due to student relocation at the end of the academic term. However, in order to substantiate that observation, a comprehensive analysis incorporating more detailed contextual information will be necessary. This work did not delve into a comprehensive examination of the underlying causes that contributed to the generation of waste that was identified. It is imperative to underscore that the socioeconomic grid data utilized in the study do not comprehensively depict the attributes of all the observed locations, including publicly owned properties and regions, thereby engendering a partiality in the suggested analysis. An instance of waste generation monitoring from a solitary waste producer location can be amalgamated with data pertaining to the surrounding populace, despite the fact that these variables do not offer a direct explanation for the waste generation at that specific location. In summary, further attention is required to address data and methodology-related concerns in the integration of container-level monitoring data and geospatial public data. One of the challenges in spatial analyses pertains to the management of spatially sparse or absent monitoring data. One may opt to utilize site-specific data that accurately represent the attributes of the monitoring location, as opposed to employing spatial analytics. The limited accessibility of such information is largely attributed to the challenges commonly associated with safeguarding data privacy and protection.

7.6 Summary

Spatio-temporal data analytics using hybrid deep belief networks (HDBNs) has emerged as a powerful tool for improving e-waste management systems. By integrating the spatial and temporal dimensions of e-waste data, these methods can help identify patterns and trends in e-waste generation, collection, and disposal. The use of HDBNs for e-waste management has been explored in several studies, which have shown that these methods can improve the accuracy of e-waste prediction models and identify key factors driving e-waste generation and disposal. These findings have important implications for policy development and decision-making in e-waste management, as they provide a better understanding of the spatial and temporal dynamics of e-waste and can help inform more effective and sustainable e-waste management strategies. Overall, spatio-temporal data analytics using HDBNs is a promising approach for addressing the challenges associated with e-waste management, and further research in this area is needed to explore its full potential. As the volume of e-waste continues to grow, it is increasingly important to develop innovative and data-driven approaches to managing this critical environmental issue, and spatio-temporal data analytics using HDBNs is one such approach that shows great promise.

References:

1. Vassakis, Konstantinos, Emmanuel Petrakis, Ioannis Kopanakis, and John Makridis. "Big data analytics on location-based social networks for knowledge generation in tourism-the case of Crete Island." *International Journal of Technology Marketing* 17, no. 1 (2023): 99-123.
2. Slavic, Giulia, Abrham Shiferaw Alemaw, Lucio Marcenaro, David Martín Gómez, and Carlo Regazzoni. "A Kalman Variational Autoencoder Model Assisted by Odometric Clustering for Video Frame Prediction and Anomaly Detection." *IEEE Transactions on Image Processing* 32 (2022): 415-429.
3. Latha, C. Jenifa, K. Kalaiselvi, S. Ramanarayan, R. Srivel, S. Vani, and T. V. M. Sairam. "Dynamic convolutional neural network based e-waste management and optimized collection planning." *Concurrency and Computation: Practice and Experience* 34, no. 17 (2022): e6941.
4. Brandt, P. C., Elena Provornikova, Stuart D. Bale, Alice Cocoros, Robert DeMajistre, Konstantinos Dialynas, Heather A. Elliott et al. "Future exploration of the outer heliosphere and very local interstellar medium by interstellar probe." *Space science reviews* 219, no. 2 (2023): 18.
5. Xu, Sa, Cunyi Yang, Zhehao Huang, and Pierre Failler. "Interaction between digital economy and environmental pollution: new evidence from a spatial perspective." *International Journal of Environmental Research and Public Health* 19, no. 9 (2022): 5074.
6. Kumar, K. Suresh, A. S. Radhamani, and T. Ananth Kumar. "Sentiment lexicon for cross-domain adaptation with multi-domain dataset in Indian languages enhanced with BERT classification model." *Journal of Intelligent & Fuzzy Systems* Preprint (2022): 1-18.
7. Xia, Jiazhi, Fenjin Ye, Fangfang Zhou, Yi Chen, and Xiaoyan Kui. "Visual identification and extraction of intrinsic axes in high-dimensional data." *IEEE Access* 7 (2019): 79565-79578.
8. Esmkhani Yuvalari, Farnaz, Bahman Zarrinjooee, and Farid Parvaneh. "Globalization of Local Lives: Performing Self-Authenticity as Personal, Local and Social Process in Zadie Smith's NW." *Critical Literary Studies* 3, no. 1 (2021): 141-156.
9. Zhu, Guangli, Zhuangzhuang Pan, Qiaoyun Wang, Shunxiang Zhang, and Kuan-Ching Li. "Building multi-subtopic Bi-level network for micro-blog hot topic based on feature Co-Occurrence and semantic community division." *Journal of Network and Computer Applications* 170 (2020): 102815.
10. Divya, P., S. Jayalakshmi, M. Pavithra, R. Rajmohan, T. Ananth Kumar, and Osamah Ibrahim Khalaf. "Accelerating graph analytics." *Demystifying Graph Data Science: Graph Algorithms, Analytics Methods, Platforms, Databases, and Use Cases* (2022): 39.
11. Namboodiri, Vijay Mohan K., and Garret D. Stuber. "The learning of prospective and retrospective cognitive maps within neural circuits." *Neuron* 109, no. 22 (2021): 3552-3575.
12. Ferreira, Leonardo N., Didier A. Vega-Oliveros, Moshé Cotacallapa, Manoel F. Cardoso, Marcos G. Quiles, Liang Zhao, and Elbert EN Macau. "Spatiotemporal data analysis with chronological networks." *Nature communications* 11, no. 1 (2020): 4036.

13. Rajmohan, R., T. Ananth Kumar, E. Golden Julie, Y. Harold Robinson, S. Vimal, Seifidine Kadry, and Ruben Gonzalez Crespo. "G-Sep: A Deep Learning Algorithm for Detection of Long-Term Sepsis Using Bidirectional Gated Recurrent Unit." *International Journal of Uncertainty, Fuzziness and Knowledge-Based Systems* 30, no. Supp01 (2022): 1-29.
14. Duan, Pinsheng, Jianliang Zhou, and Yang Miang Goh. "Spatial-temporal analysis of safety risks in trajectories of construction workers based on complex network theory." *Advanced Engineering Informatics* 56 (2023): 101990.
15. Arifin, Sutjipto, and Peter YK Cheung. "A computation method for video segmentation utilizing the pleasure-arousal-dominance emotional information." In *Proceedings of the 15th ACM international conference on multimedia*, pp. 68-77. 2007.
16. Arain, A. L., R. Pummill, J. Adu-Brimpong, S. Becker, M. Green, M. Ilardi, E. Van Dam, and R. L. Neitzel. "Analysis of e-waste recycling behavior based on survey at a Midwestern US University." *Waste Management* 105 (2020): 119-127.
17. Moyle, Brent D., David B. Weaver, Stefan Gössling, Char-lee McLennan, and Arghavan Hadinejad. "Are water-centric themes in sustainable tourism research congruent with the UN Sustainable Development Goals?." *Journal of Sustainable Tourism* 30, no. 8 (2022): 1821-1836.
18. Suresh Kumar, K., and C. Helen Sulochana. "Local search five-element cycle optimized reLU-BiLSTM for multilingual aspect-based text classification." *Concurrency and Computation: Practice and Experience* 34, no. 28 (2022): e7374.
19. Plotnikova, Veronika, Marlon Dumas, and Fredrik P. Milani. "Data mining methodologies in the banking domain: A systematic literature review." In *Perspectives in Business Informatics Research: 18th International Conference, BIR 2019, Katowice, Poland, September 23–25, 2019, Proceedings 18*, pp. 104-118. Springer International Publishing, 2019.
20. Kerdlap, Piya, Jonathan Sze Choong Low, and Seeram Ramakrishna. "Zero waste manufacturing: A framework and review of technology, research, and implementation barriers for enabling a circular economy transition in Singapore." *Resources, conservation and recycling* 151 (2019): 104438.
21. Hyvönen, Eero. "Using the Semantic Web in digital humanities: Shift from data publishing to data-analysis and serendipitous knowledge discovery." *Semantic Web* 11, no. 1 (2020): 187-193.

Chapter 8
Spatiotemporal and Intelligent Transportation Forecasting

K. Maithili, S. Leelavathy, G. Karthi, and M. Adimoolam

Abstract Spatiotemporal-based intelligent transportation systems are increasingly being integrated into various surveillance systems. To enhance the efficiency of these systems, automated forecasting was introduced to identify and penalize non-compliant behaviors. This chapter explores a range of location-based transportation forecasting systems and the necessary adaptations for smart cities. Additionally, the frameworks of transportation systems using intelligent methods have been evaluated to analyze their merits and demerits. Subsequently, route-based prediction was examined for its real-time application efficacy. Building upon spatial forecasting methods, their essential techniques and adaptation potentials have been explored for the transportation application of spatiotemporal data. In conclusion, analysis and forecasting-based performance metrics are presented, focusing on intelligent transportation systems across various countries, along with their challenges. As a key focus, the applications of spatiotemporal-based intelligent transportation forecasting are discussed in detail.

K. Maithili
Department of Computer Science and Engineering, KG Reddy College of Engineering and Technology, Hyderabad, Telangana, India

S. Leelavathy
Department of Computer Science and Engineering, Aarupadai Veedu Institute of Technology, Vinayaka Mission Research Foundation, Chennai, Tamil Nadu, India
e-mail: leelavathy@avit.ac.in

G. Karthi
Department of Artificial Intelligence and Data Science, Saveetha Engineering College, Chennai, Tamil Nadu, India
e-mail: gkarthi@saveetha.ac.in

M. Adimoolam (✉)
Department of Computer Engineering, Government Polytechnic College, Tiruvallur, Tamil Nadu, India
e-mail: adimoolam@gptcsevvapet.co.in

J. A et al. (eds.), *Spatiotemporal Data Analytics and Modeling*, Big Data Management, https://doi.org/10.1007/978-981-99-9651-3_8

Keywords Transportation system · Intelligent transportation system · Forecasting the intelligence of transportation · Route-based prediction · Spatial forecasting method

8.1 Introduction to Location-Based Transportation Forecasting

In an era of technological advancement, well-versed nations that are establishing modernized cities have introduced smart transportation systems to mitigate traffic congestion and optimize routes, time, and energy consumption [1]. Simultaneously, this smart transportation framework has incorporated advanced Machine Learning and Deep Learning algorithms to enhance its efficiency. Typically, location- or spatial-based transportation forecasting methods are crucial in sophisticated transportation systems. Various accurate traffic prediction methods are being utilized and developed, particularly for forecasting traffic transportation, including time series forecasting models. The advantage of time series forecasting is its flexibility, allowing predictions to be tailored for specific times, such as daily morning and evening peak hours, as well as weekly and monthly patterns.

Further traffic forecasting is supposed to meet the challenges of proportionate transportation alongside intelligence since weather conditions of the road also decide the forecasting difficulty. A work was introduced to handle this situation: Attribute-Augmented Spatiotemporal Graph Convolutional Network (ASTGCN). ASTGCN used the real dataset and showed the external factors, including static and dynamic spatiotemporal information. The traffic speed is predicted with forecasting and is compared with traditional traffic prediction methods. The accuracy performance was compared to existing methods for transport prediction forecasting [2] (Fig. 8.1).

8.1.1 ML-Based Models for Transportation Forecasting

For transportation traffic prediction, ML- and DL-based forecasting models were used by smart cities of many nations. Some ML models majorly used are Kalman Filtering (KF), Auto-Regressive Integrated Moving Average (ARIMA), Holt-Winters' Exponential Smoothing (HWES), Random Walk (RW) model, K-Nearest Neighbor (KNN) Algorithm [3], and K-Dimensional Tree (KDT). These models are categorized as time series transportation forecasting. Table 8.1 lists the ML methods alongside traffic applications with their drawbacks. These algorithms are used to predict short-time traffic [4, 5], travel time prediction, road traffic prediction, and network traffic prediction [6].

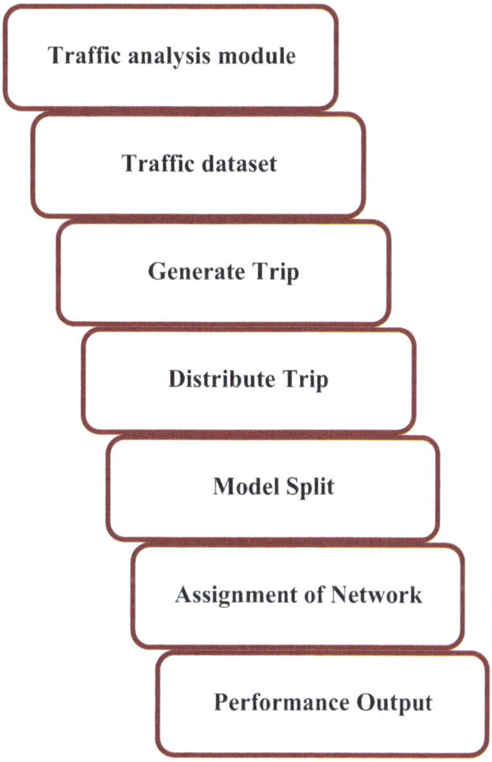

Fig. 8.1 Traffic analysis steps for traffic prediction

Table 8.1 Comparison of ML-based time series traffic prediction's transportation forecasting methods

Method	Traffic-based applications	Drawback
KF	Traffic with short time	Unsatisfied accuracy
ARIMA	Travel time prediction, road traffic prediction	Large size data required
HWES	Network traffic modeling	Vast seasonal traffic prediction
KNN	Travel time prediction	Confusion with redundant features
KDT	Road traffic prediction	Nonlinear precision
RW	Time-based traffic prediction	More complex processing steps

8.1.2 DL-Based Models Used Time Series Transportation Forecasting

Further DL-based models were also used for transportation forecasting with time series concerns. This model for transportation forecasting might not only have some

Table 8.2 Comparison of DL-based time series traffic prediction's transportation forecasting methods

DL method	Traffic-based applications	Drawback
RNN	Traffic time prediction	More memory space used
CNN	Road traffic prediction	Complex structure
STL	Travel time prediction, road traffic prediction	Small dataset oriented
SMOTE	Time-based traffic prediction	Time increases with dataset volume
LSTM	Traffic with short time	Memory and time consumption are more

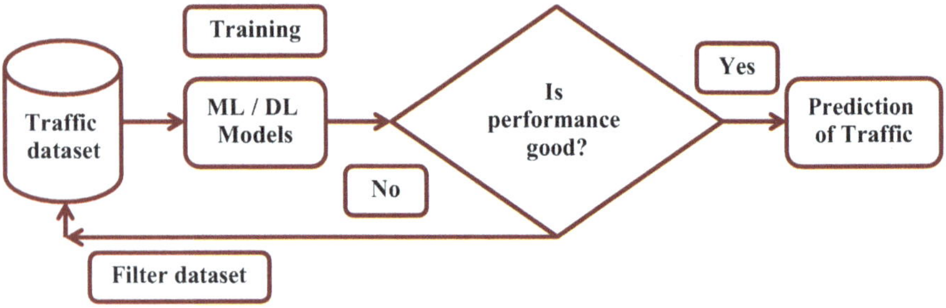

Fig. 8.2 General ML- and DL-based training model for traffic prediction

benefits but also challenges in predicting traffic due to low stability and high temporal dependencies. Also, DL-based traffic forecasting needs to improve adaptability [7]. To gain and extend prediction, sometimes a hybrid DL was used. The major DL algorithms were Convolutional Neural Network (CNN) [8], Long Short-Term Memory (LSTM), Synthetic Minority Oversampling Technique (SMOTE), Recurrent Neural Network (RNN) [9], and SMOTE and Tomek Links (STL) [10]. Table 8.2 lists the various DL-based methods and traffic-based applications with their demerits. All these methods are used to predict traffic speed, traffic flow, traffic travel time prediction, short-time traffic prediction, etc. [11].

8.1.3 Prediction of Traffic with ML and DL Model

There are numerous established methods utilizing Machine Learning (ML) and Deep Learning (DL) that can effectively predict traffic patterns using datasets. As illustrated in Fig. 8.2, the traffic prediction process and its evaluation are systematically outlined. A variety of ML and DL models, including Support Vector Machine (SVM), KNN, ARIMA, KF, AWES, RW, and KDT [12–14], have been employed for the classification and detection of network traffic. In a typical implementation of these models, traffic datasets are first trained using a selected ML or DL model. The model's performance is then meticulously evaluated for precision. If the model achieves satisfactory predictive accuracy, it indicates the success of the traffic prediction. Conversely, if the results are suboptimal, feedback is directed back to the dataset for refinement. This involves filtering and improving the dataset before reapplying the same model for enhanced traffic prediction accuracy.

8.1.4 Required Components for Transportation Forecasting

For traffic transportation forecasting, various components were used: the GPS and its data, smart card, traffic data source sensors, call details record, mobile phone, etc.

GPS and Its Data: GPS serves as a pivotal tool in gathering transportation data, transforming each equipped vehicle into a mobile data collector. It provides crucial temporal and spatial information about vehicles, essential for traffic forecasting. Key features and parameters like distance, trip time, and travel speed are recorded with an emphasis on minimizing reporting burdens. Different GPS versions, such as logging GPS and real-time tracking GPS, are evaluated for their advantages and limitations in this context.

Smart Card of Traffic Handler: This technology enables the analysis of individual travel patterns using specific transportation tools. It examines parameters such as travel direction, start and end times of journeys, and frequently visited destinations. By analyzing these individual parameters, traffic patterns can be effectively discerned, guided by a traffic handler system. The primary challenge here lies in addressing the heterogeneity of the data.

Traffic Data Source Sensors: "Sensors" in this context refer to devices and components installed at smart transportation sites along highways to capture real-world data. Common measurements include hourly traffic volume, average vehicle speed, and vehicle occupancy. These data points are instrumental in developing efficient modules and models for smart transportation systems.

Call Detail Record (CDR): Initially used for billing purposes, CDR has evolved into a tool for gathering longitudinal and latitudinal data of users in transportation systems, complemented by timestamp analysis. CDR enables more effective and regular analysis than traditional surveys and is instrumental in developing precise transportation models using captured individual route information.

Mobile Phone: Mobile phone is the key component in traffic transportation prediction systems. In the Vehicle Ad-hoc Network (VANET), mobile will act as a vehicle segment for data transmission, especially for transit trips. Mobile phones are used to provide user location pictures with accuracy and precision. This data also reduces the burden and location-based users' details for a long time. Further, this segment also introduced high-resolution travel path images alongside all users' out-vehicle and in-vehicle trip details. The high challenge over this mobile data of users' location is that it is a high volume of data, and big data processing will be tough. Storage and power to process data will be highly overburdened [15].

Connected Vehicle (CV): It is a smart transportation to establish the next generation. This sophisticated technology can introduce green transportation alongside intelligent transportation for the modernized world and certainly impact society's adaptation to new transportation optimization-related things. A major concern of this connected vehicle can introduce transported safety and secure applications. The essential component of this CV is that I can introduce the Internet of Vehicles (IoV) from the modified version of the Internet of Things (IoT). The IoV can be able to produce reliable as well as efficient autonomous vehicle comfort for safe mobility. This CV can be extended and exposed to Vehicle to X (V2X) for the interaction of smart vehicles. Further, it was categorized as Vehicle to Vehicle (VoV), Vehicle to Sensors (VoS), Vehicle to Internet (VoI), and Vehicle to Road (VoR). The communication components used for this extended transportation technology are Ultra Wideband and Bluetooth [16].

8.2 Framework for Intelligent Transportation System

In this section, we delve into various model-based frameworks for transportation systems, with a specific focus on the use of Neural Networks (NN) in Computer-Aided Detection (CAD). The primary goal of employing NN in CAD is to enhance classification accuracy and reduce inference time. A cutting-edge method, known as Convolutional Neural Network (CNN), has been proposed due to its promising potential.

Drawing inspiration from the vision system of cats, which is orientation-selective and regionally sensitive, CNNs utilize fully connected layers with shared weights. This approach allows for a comprehensive analysis of the entire image by extracting local features using a single kernel. The proposed detection methodology boasts superior accuracy and reduced memory requirements compared to conventional techniques.

Unlike traditional methods that rely on handcrafted feature extraction, CNNs offer an automated learning mechanism through back-propagation. This process facilitates the hierarchical extraction of spatial features across the network. The structure of this model is schematically represented in Fig. 8.3. It comprises layers such as convolution, pooling, and fully connected layers, which are integral to the

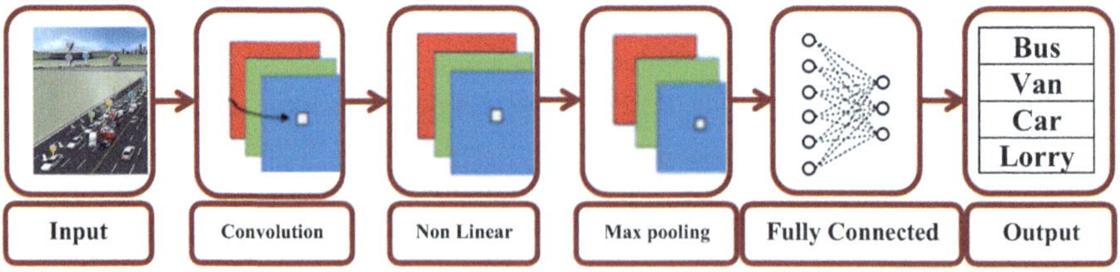

Fig. 8.3 CNN-based framework for intelligent traffic prediction system

network. The first two layers primarily focus on feature extraction, with the final layer serving as a classifier.

The convolutional layer employs a convolution kernel, which is pivotal in activating and computing functions. Notably, the activation functions in this layer are inherently nonlinear, yet the processing and phasing of the layers follow a linear pattern. The kernel, utilizing the concept of shareability, is instrumental in extracting neighboring information, a process considered a fundamental gain in this context. It adeptly captures spatial hierarchical information through the activation function. In this framework, parameter values adjust in response to neighboring inputs. The mapping of these parameters is executed via a nonlinear activation function, reflecting neural characteristics, and ensuring a one-to-one ("onto") function. Following the convolutional layer is the pooling layer, typically engaged to handle a range of parameters and identify their maximum values. This layer is critical in down sampling processes, employing either average or max-pooling techniques based on the specific requirements that arise. The subsequent processing, whether through max-pooling or activation functions in the convolutional layer, leads to feature extraction. The network then maps these extracted features to the total weight in the fully connected (linked) layer of the CNN, further facilitating down sampling. Before the feedforward operation, statistical evaluations are conducted. In certain cases, the output from the max pooling, fully connected, and convolutional layers might appear identical, leading to the initiation of the classification phase. During classification, various likelihood classes are compared, and immediate results are obtained through an activation function with a preset checking threshold. The sigmoidal function is often employed here for binary and multi-class classification, while a regression function is invoked to refine the output. It is noteworthy that the widespread adoption of CNNs and similar models gained momentum with the availability of large volumes of labeled data, particularly in initiatives like the ImageNet challenge [17].

8.3 Route-Based Prediction with Intelligent Transportation Systems (ITS)

In this research, advanced deep learning strategies, guided by Intelligent Transportation Systems (ITS), have been deployed for predicting efficient and intelligent routes. The analysis extends to previously unexplored statistical dimensions within transportation domains. Additionally, enhancements in network traffic management focus on administration, planning, and safeguarding sensitive information. A significant aspect of this study is the testing of protected network traffic in road transport scenarios, emphasizing cost reduction in both public and private sectors. This also includes shared transportation experiences. Autonomous vehicles were subjected to rigorous testing, demonstrating a new benchmark in traffic prediction systems for road transport. This work presents a comprehensive approach to traffic perception, enriched through test drives and an evaluation of the transportation infrastructure using sophisticated deep learning algorithms. The core of the methods discussed here revolves around ITS-integrated deep learning algorithms. These "ultra-model" methods were meticulously evaluated for their performance within an ITS-based deep learning framework. Further in-depth analyses included the classification of vehicles during the training phase of the deep learning algorithm. This contemporary classification approach significantly contributes to the strategy employed in transportation traffic systems. The deep learning algorithm, enriched with ITS-informed insights, was retrained to further refine its efficacy in managing transportation traffic, with a particular focus on expanding the traffic dataset size to gauge the system's capacity. The datasets, meticulously labeled, underwent training using supervised learning algorithms. The culmination of this process involved feeding embedded, specific transportation traffic features into the deep learning model for ITS deployment. Subsequently, a thorough discussion on the merits and demerits of these methods was undertaken, based on the tested inputs and corresponding outputs [12].

8.3.1 RNN-Based Architecture

In this study, we employed a supervised deep learning (DL) approach, utilizing Recurrent Neural Networks (RNNs), to analyze dynamic datasets characteristic of data sequences. The RNN model facilitates the forward feeding of data interpretations for assessing statistical sequences in performance parameters. These parameters are rigorously evaluated through test-driven training within transportation traffic systems, ensuring the integrity and reliability of the input data. A particular focus was placed on underpass data, chosen for its relevance in evaluating throughput performance in transportation traffic systems. This approach involves a stepwise procedure to measure performance metrics. A selective dataset is analyzed to ascertain outputs, particularly focusing on the accuracy of transportation traffic predictions. Subsequently, performance times are examined, including the evaluation of newly entered vehicle data, analyzed against both post-metric and preceding traffic times. The study also includes a comparative analysis, measuring the similarity index between traditional algorithms, the Markov model, and RNN. Here, the RNN, operating on a recurrent network basis, differs fundamentally from the Hopfield neural network structure. For simplification, the RNN is referred to as "simplified RNN" in this context. Figure 8.4 delineates the schematic architecture of both the RNN and the Markov model, highlighting their roles in the optimization of classified objects. The simple RNN diagram presents a modified version of the RNN, positioned centrally in the structure. Initially, the Hopfield network's efficacy in sample recognition was assessed as a preliminary test.

Although an RNN network appears straightforward, the two fundamental difficulties remain vanishing or exploding gradient issues. These troubles can appear at some stage in getting to know from preceding states when the chain of dependencies

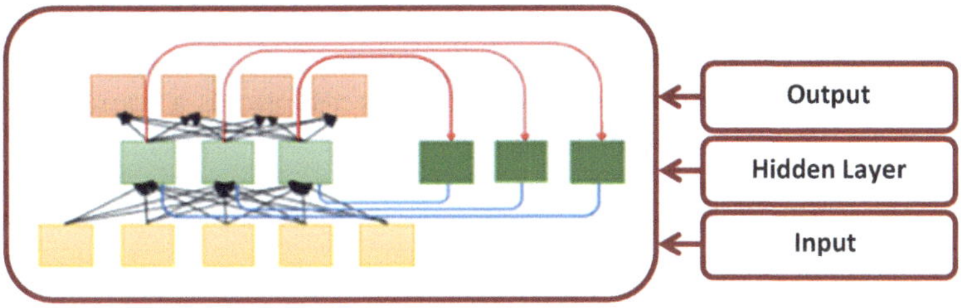

Fig. 8.4 RNN-based framework for intelligent transportation

is extended, and, in this case, it is tough to select which data have to be realized from previous states [18, 19]. Recurrent networks used an exploding gradient structure to resolve the trouble in the processed input dataset. Since oscillating weights and their results are colliding sometimes. The reasons are Truncated Back-Propagation Through Time (TBPTT) network architecture for transportation traffic prediction systems. The steps used to check the certainty are propagation steps of number units. Sometimes the small components of RNN were tested for past analysis under statistical information gathering. The coaching phases of RNN are very simple and disturbed with exploited gradient descent parsing. However, in the case of future cravings belongings, the previous measurements connected with these cravings will abandoned up lost.

To clear up each of these troubles introduced by RNN for transportation traffic, the LSTM structure has been recommended. The main concept of this technique is the usage of a reminiscence telephone with solely double openings of enters and output. The middle opening evaluates, how to preserve the statistics in the phone, and the harvest opening resolves when to get the right of entry to the reminiscence cellphone or stop its impact on different units. More than a few rectifications and upgrades were used to LSTM architecture in topical years.

Usually, LSTM carries a recollection handset, and it has tightened up its nation's prolonged series of sequences and created totally on its parameters. It controls the handset to influence the network. Even though the LSTM mobile phone has used a variety of architecture, this version was suggested. Several gates and aspects introduced to this telephone are distinct from the fundamental suggested. For gate activation, the logistic sigmoid feature is used for common feature detection. The sketch and a Tanh feature are used to block the center for input feed and later output is activated. The neglect gate and peephole connections have been first advised that allows the telephone to reset by forgetting its contemporary nation and passing the cutting-edge nation statistics from the interior country to all openings barring short-lived on LSTM through an initiation occupation.

The Gated Recurrent Unit (GRU) was introduced finally for stimulation, and it was used as the LSTM block. The pipe feed peephole connections were used to test the input and output stimulation gathering performed later, and it was eradicated. The coupled feature is tested for entrance, and it was used as the neglect gates to convert into a reference unit. Later, it was replaced with an input gate. The passes are trained with recurrent connections, and it acts as the block input. Since the LSTM is introduced which are reduced intro linear architecture it has a simple structure and elimination can be carried out. This also avoids a considerable discount in routine, making it extra famous for use.

8.4 Spatial Forecasting Methods

The governments and some transportation companies required the improvement of the Internet and function technologies. Further, there are extra and extra spatiotemporal facts accumulated and thereby some forecasting methods are required over spatial data. The spatial data usable companies have used some strategies for gathering corresponding trajectories. The orders are able to be completed with the respective area. An example of such a transportation forecasting company is Uber Footnote which deals with 30 million and 18 million journey orders daily. Using the accumulated records to enhance troubles and deliver traffic handy conveyance offerings to people is herbal. In different disputes, achievements are to be made transference sensible on or after accrued spatiotemporal records. Solitary major technique to advance the purpose is primarily created on predicting traffic transport and using spatiotemporal data. The interest in each educational uses traffic prediction, and it has the hassle of attracting public transport. Once again many industries in traffic hassle and neck moments. The intelligence systems are used to predict the large dataset for spatial data forecasting, and it has a spectrum of extension over the given work and proposed models. Further this work has initiated a prediction system for problem feeding and output gathering. An intelligent transportation system has been initiated for these transportation applications.

Figure 8.5 depicts the spatial data-based model with their steps for traffic prediction general working mechanism. The initial step of this spatial data forecasting is dataset collection. This dataset may have various formats like temporal data, spatial data, static spatiotemporal data, or dynamic spatiotemporal data. The next step will be the processes: pre-processing of the dataset, data cleaning on the dataset, data compression of the dataset and finally, the data storage for a dataset of spatial

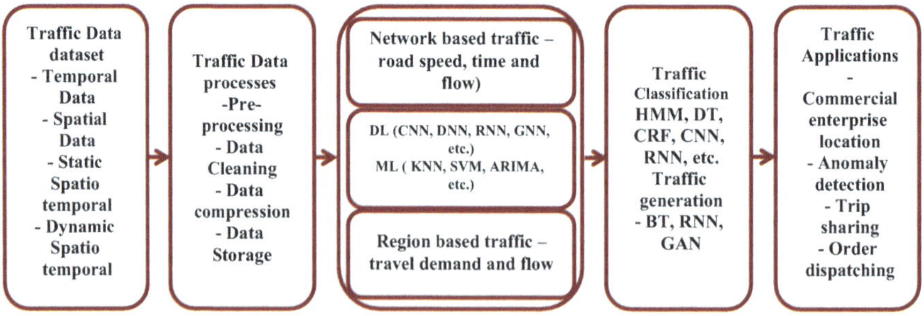

Fig. 8.5 Methods of spatial data forecasting

data. The third step is the heart of this spatial data forecasting. This forecasting may be followed differently like network and region-based traffic forecasting. In network-based traffic forecasting, traffic applications such as road speed measurement [20], traffic time prediction, and traffic flow prediction can be carried out. With region-based traffic prediction, traffic applications like travel demand and traffic flow prediction can be carried out. Among networked and region-based traffic forecasting prediction, the traffic flow prediction can be performed as a common challenge, as illustrated in the middle block of Fig. 8.5. The fourth step in spatial forecasting is classifying traffic and traffic generation. The ML and DL methods like HMM, DT, CRF, CNN, RNN, etc. are used to classify traffic forecasting. The ML and DL methods, such as BT, RNN, and GAN, are used in traffic generation. Finally, the step is utilization for traffic applications. The famous traffic forecasting of transportation is the identification of commercial enterprises, Anomaly detection on traffic prediction, trip sharing prediction and, which was majorly used by many commercial purpose applications. Order dispatching is another application of traffic transportation.

Prediction of Traffic Status: Navigation devices are used majorly in digital mapping for travel route finding including shortest route and optimal route based on user experience. It will avoid traffic congestion and suggest a good route based on current traffic jams and rerouting information. Traffic vicinity is confidential for traffic prediction and traffic rerouting. Here the goal was tested with the trained capability of the model, and their goal was tested further over the congested road traffic dataset. After the test, the trained dataset of road traffic was tested again with almost all routes of road traffic. The traffic status is covered with travel time and tested for slower traffic speed. The performance measures tested here are trained with roads like smooth, heavily congested roads as well as mediocre road-based traffic. The traffic repudiation is measured with linear regression analysis and prediction time as noted to plot the graphical representation of the output. Again the non-stop interval set to measure the performance and prediction was calculated for further process.

Prediction of Travel Demand: Online taxi providers are used to check the requirement of transport for commercial as well as individual users [21]. Usually, tourist routes and their travel demands are measured, and it is essential to provide hassle-free travel offers and comfortless. Usually, the residential areas can be utilized for travel routes in peak hours than commercial routes. But sometimes it will be reversed due to unavoidable situations. On the other hand, motor vehicles are booked for popular routes with highly impacted routes with any kind of traffic and even in peak hours. This will contrast the proper road transport traffic and predicting the journey place are very effective with this model and algorithms.

Prediction of Traffic Flow: Prediction of traffic flow is an essential task since the free route and busy routes are to be checked before progress the vehicle over the route. On the other hand, it is important to provide sophisticated route information, and it can only be possible to predict the traffic flow. But some time some immoderate as well as stomp traffic occasion come to traffic area [22]. To predict the traffic flow, entire traffic dataset are region-based network traffic and network-based network traffic. The traffic flows are predicted as per following scheduled practices. For example, if a hundred human beings are leaving place A for the vicinity B, each A's out-flow and B's in-flow would make one hundred bigger.

8.5 Forecasting and Analysis of Metrics

For forecasting transportation, performance measures such as Mean Absolute Percentage Error (MAPE), Route Mean Square Error (RMSE), and Standard Errors (SE) can be used [23]. The RMSE is one of the important overall performance symptoms for a regression mannequin for traffic network transport prediction. It measures the common distinction between site visitors of transport values expected by way of an ML or DL mannequin and the real values of transport traffic. It estimates how properly the ML or DL mannequin can predict the goal price transport traffic as represented in Eq. (8.1) [24].

$$\mathrm{RMSE} = \frac{\sqrt{\sum_{i=1}^{N}\left(\mathrm{Predicted}\,i - \mathrm{Acutal}\,i\right)}}{N} \tag{8.1}$$

MAPE is a relative error measure of traffic transportation that makes use of absolute values of transport visitors to preserve the practical and bad blunders from cancelling one another out and makes use of relative mistakes of transport visitors to allow you to examine forecast accuracy between time series fashions for transport traffic. It is represented in Eq. (8.2).

$$\mathrm{MAPE} = 100\% / n\sum_{t=1}^{n} \#At - Ft / At\# \tag{8.2}$$

The traffic of Transportation Sampling is a traffic prediction system associated with traffic observation resolution; it helps us make statistical inferences about the traffic population. Accumulating all the facts about the traffic pattern is time-consuming and expensive. Therefore, traffic sampling is done to make inferences about the traffic population. In sampling, traffic pattern samples are expected to be drawn from the traffic populace, and pattern ability and populace capability are equal to traffic patterns. Given that a traffic pattern is a crew of character observations drawn from a traffic population, the range of character devices of traffic or observations representing a pattern is recognized as the pattern size of traffic. Since the pattern intends to furnish a consultant pattern of the whole population over transportation traffic, the larger the pattern size, the more correct the dimension and the greater the self-assurance of the traffic pattern. It has that the pattern is consultant of the total population of a pattern of traffic.

Let's say SE has a distribution of non-stop random variables. From the distribution, SE takes a pattern of 10 observations and plots the commonality of the N

observations. SE repeats this workout quite a few times. On every event, SE accumulates a pattern of *N* observations from the distribution and plots the commonality of the *N* observations. As SE preserve repeating this exercise, the sampling distribution of the pattern will signify an everyday distribution. If SE had been to repeat the whole workout with large pattern sizes of the transportation traffic, the curve would emerge as narrower and taller, and the popular deviation would be tighter transportation traffic. SE will word that with large pattern sizes of prediction of traffic transportation, the pattern of traffic implies will strategy the populaces mean for traffic. The SE tells you how your pattern suggests deviating from what the populace means. The larger your pattern size, the nearer your pattern of traffic prediction implies to the true populace mean transportation of traffic. Note that the SE is very comparable to the Standard Deviation. Both are measures of spread. The greater the number, the more your information unfolds on traffic prediction of transportation.

However, there is a key distinction between the SE and Standard Deviation. The preferred error uses data on traffic of transportation (i.e., pattern data), whereas the fashionable deviation uses populace data on traffic transportation. The general deviation of the sampling distribution of the pattern suggests, additionally known as the trendy error of suggest. It is computed as follows for traffic transportation: trendy deviation of the sampling distribution of pattern suggests/rectangular root of pattern size of prediction of the traffic of transportation, and it is represented in Eq. (8.3).

$$\mathrm{SE} = \frac{\sigma}{\sqrt{n}} \qquad \sigma = \sqrt{\sum_{i=1}^{n} (yi - \mu i) / n - 1} \tag{8.3}$$

Where *yi'* is the predicted transportation value and *yi* is the actual transportation value, σ and μy are the mean transportation value.

8.6 Spatiotemporal-Based Intelligent Transportation Forecasting Applications

The traffic forecasting viable to reap sensible transportation, many purposes have to be developed based totally on the prediction of traffic [25]. In this section, some commonly used applications of traffic surveyed, which are, respectively, referred to as commercial enterprise location, anomaly detection, trip sharing, and order dispatching. In calculation, these presentations closely count the number preceding the overall routine of stream of traffic prophecy procedures and methods. For illustration, earlier than transmitting minicab briefings, figuring out trip sharing practices or arrangement directions for manipulators, traffic applications ought to approximation the journey time or passenger pace on avenue networks. On the other hand, the greater the correct expected forthcoming stream of traffic positions, the higher the stream of traffic software gas station. Now it is integral to explain briefly.

8.6.1 Commercial Enterprise Location

Commercial enterprise location is usually used to find the tour route information and the impact is that the tour route should be very short and comfortable [26]. It has leveraged historic trajectory facts and spatial aspects for the route of tourist places. Finally, it has introduced a framework for tour deployment under commercial enterprise locations. A technique was used to formalize the traffic trouble, and it was Integer Linear Programming (ILP). This optimization technique is able to consider NP-hard too. A multiple-objective optimization problem was discussed and their hassle-free idea was initiated [27].

Selecting the most desirable areas to area retail stores, charging stations or billboards can extend commercial enterprise turnover. Unsurprisingly, stream traffic records such as rush-hour traffic network streams and tour spell performance are measured to a great position in fixing this delinquent. A work was attempted to initiate the settlement of superior work [28] and it has a location-based social network alongside recognition of objects attempt to discover the superior settlement of merchandising keep with location-based social networks. This work identified objects which are used to identify locations to install retail shops and their reputation details. It has studied the geographic as well as mobility nature of fashion shops and network flow. Finally, it advises the installation of commercial enterprises and featurism places.

8.6.2 Anomaly Detection over Spatial Temporal Processing

Yet another application of traffic transportation of spatiotemporal processing is as follows. Differently, Zhang et al. [29] put into effect Deep Trust Network (DBN) and LSTM fashions. It could observe the tournament tweets associated with network-based traffic accidents. This idea was initiated to measure the totality of social media data. A learning technique alongside the relationship between traffic accidents and human mobility was discussed [30]. A denoise autoencoder mannequin technique was used to research hierarchy, and it has introduced function illustration of human mobility. Further, this mobility was used to predict traffic accident threat levels in all aspects. The influences are initiated by auto accidents and sports activities in a spatial data fashion. Further, these processes are extended with video games and concerts [31]. Later work was used by the CNN mannequin to observe traffic accidents, and it is purely based totally on traffic flows [32, 33].

A work was introduced to partition every trajectory [34]. This was used to set line segments for traffic indication. Later the anomaly was noticed on computation similarity between specific units of line segments. Later work used anomaly trajectories for detection to assist in crook behaviors of taxi drivers as discovery [35, 36].

8.6.3 Trip Sharing

Trip sharing nowadays is a demand system since the consumption of fuels for sophisticated travel can be shared and thereby the travel cost can be shared among travelers. Also, the income can be maximized and on the other hand, a friendly environment can be made. Trip sharing also helps the transport people wealthy sometimes. Thereby the travel time is also reduced with optimal cost, optimal vehicle usage optimal distance, and a friendly environment. Finally, transportation traffic also can be reduced considerably. More and more human beings are excellent at sharing their journey with others due to the full use of assets and environmental friendliness. The concept of trip sharing is that a common vehicle can pick up multiple passengers at a linear distance and optimized fashion [37]. The sharing cost is also divided accordingly anywhere and anytime [38–40]. Virtual reality and augmented reality play tremendous roles in the sharing concept since every passenger who wants to travel in trip share can able to know the vehicle movement and in that gap of distance, upcoming travelers can wait or do work till the vehicle approaches them [41], the data structure used here are Kinetic Tree (KT) to enhance the efficiency among passengers. Casual things can be done here for trip sharing effectively and efficiently [42].

Some other work used trip sharing and the techniques were dynamic mechanisms of the Branch and Bound (BB) method [43]. Again the random mixture concept like the Integer Linear Packages (ILP) method is also used. Both are well-versed in batch orders after individual orders of taxis [44, 45].

8.6.4 Order Dispatching

Order dispatching is a very interesting and compact concept nowadays used in transportation systems, and it has effectively delivered the ordered items appropriately and in a timely manner [46]. This mechanism used near and bound mechanisms to collectively deliver the ordered items on a first come first served passion. Here ML algorithms like reinforcement learning and rule-based learning are effectively applied [47, 48]. Here quality matching was preferred to deliver the order as dispatching systems. Sometimes the coupled passion was used since two orders of different shops can be delivered by a single person after a combined order. This concept can enhance delivery with reduced transport costs and also gain the benefits of delivery personnel [48–50].

8.7 Summary

In this chapter, Spatiotemporal and Intelligent Transportation Forecasting for various vehicle surveillance systems are discussed. Later, their related concept was discussed to address the forecasting for intelligent transportation systems. Initially, various location-based transportation forecasting systems and adaptations necessary for smart cities are discussed. Further, the frameworks of transportation systems with intelligent methods have been studied and their merits and demerits. Later route-based prediction was analyzed for real-time applications and briefly technically. The forecasting and their necessary techniques and adapting possibilities have been analyzed for transportation application of spatiotemporal data. Next, the forecasting-based performance metrics have been viewed towards intelligent transportation systems for various countries and their challenges are discussed elaborately with their equations. Finally, the spatiotemporal-based Intelligent Transportation forecasting applications are discussed elaborately in the concluding part of this book chapter.

References

1. Zahra Karami, Rasha Kashef, Smart transportation planning: Data, models, and algorithms, Transportation Engineering, vol. 2, pp. 100013, 2020. https://doi.org/10.1016/j.treng.2020.100013.
2. J. Zhu, Q. Wang, C. Tao, H. Deng, L. Zhao and H. Li, "AST-GCN: Attribute-Augmented Spatiotemporal Graph Convolutional Network for Traffic Forecasting," IEEE Access, vol. 9, pp. 35973-35983, 2021, https://doi.org/10.1109/ACCESS.2021.3062114.
3. Dairi A, Harrou F, Sun Y, Senouci M (2018) Obstacle detection for intelligent transportation systems using deep stacked autoencoder and k-nearest neighbor scheme. IEEE Sens J 18:5122–5132
4. Liu Y, Wang Y, Yang X, Zhang L (2017) Short-term travel time prediction by deep learning: a comparison of different LSTM-DNN models. In: 2017 IEEE 20th international conference on intelligent transportation systems (ITSC), IEEE, pp 1–8.
5. Yao B, Chen C, Cao Q, Jin L, Zhang M, Zhu H, Yu B (2017) Short-term traffic speed prediction for an urban corridor. Comput-Aid Civ Infrastruct Eng 32(2):154–169. https://doi.org/10.1111/mice.12221
6. Gang X, Kang W, Wang F, Zhu F, Lv Y, Dong X, Riekki J, Pirttikangas S (2015) Continuous travel time prediction for transit signal priority based on a deep network. In: 2015 IEEE 18th international conference on intelligent transportation systems, IEEE, pp 523–528.
7. Wang Y, Zhang D, Liu Y, Dai B, Lee LH (2018a) Enhancing transportation systems via deep learning: a survey. Transp Res Part C Emerg Technol 99:144–163
8. Li L, Qu X, Zhang J, Wang Y, Ran B (2019) Traffic speed prediction for intelligent transportation system based on a deep feature fusion model. J Intell Transp Syst 23(6):605–616. https://doi.org/10.1080/15472450.2019.1583965
9. Fu X, Luo W, Xu C, Zhao X (2020) Short-term traffic speed prediction method for urban road sections based on wavelet transform and gated recurrent unit. Math Probl Eng 2020:1–13. https://doi.org/10.1155/2020/3697625
10. Gu Y, Lu W, Qin L, Li M, Shao Z (2019) Short-term prediction of lane-level traffic speeds: A fusion deep learning model, vol 106. https://doi.org/10.1016/j.trc.2019.07.003.

11. Liu D, Tang L, Shen G, Han X (2019) Traffic speed prediction: An attention-based method. Sensors 19:3836. https://doi.org/10.3390/s19183836
12. Haghighat, A.K., Ravichandra-Mouli, V., Chakraborty, P. et al. Applications of Deep Learning in Intelligent Transportation Systems. J. Big Data Anal. Transp. 2, 115–145 (2020). https://doi.org/10.1007/s42421-020-00020-1
13. Xu, C., Zhang, A., Xu, C. et al. Traffic speed prediction: spatiotemporal convolution network based on long-term, short-term and spatial features. Appl Intell 52, 2224–2242 (2022). https://doi.org/10.1007/s10489-021-02461-9
14. Yuan, H., Li, G. A Survey of Traffic Prediction: from Spatio-Temporal Data to Intelligent Transportation. *Data Sci. Eng.* **6**, 63–85 (2021). https://doi.org/10.1007/s41019-020-00151-z
15. Zhu L, Yu FR, Wang Y, Ning B, Tang T (2018a) Big data analytics in intelligent transportation systems: a survey. IEEE Trans Intell Transp Syst 20(1):383–398
16. Niu K, Zhang H, Zhou T, Cheng C, Wang C (2019) A novel spatio-temporal model for city-scale traffic speed prediction. IEEE Access 7:30050–30057
17. Zheng C, Fan X, Wang C, Qi J (2020) Gman: A graph multi-attention network for traffic prediction. In: AAAI. pp. 1234–1241
18. Liu Y, Liu Z, Vu HL, Lyu C (2020) A spatio-temporal ensemble method for large-scale traffic state prediction. Comput-Aid Civ Infrastruct Eng 35(1):26–44. https://doi.org/10.1111/mice.12459
19. Xu DW, Wang YD, Jia LM, Qin Y, Dong HH (2017) Real-time road traffic state prediction based on arima and kalman filter. Front Inf Technol Electron Eng 18 (2):287–302. https://doi.org/10.1631/FITEE.1500381
20. Cui Z, Zhao C (2019) Spatio-temporal broad learning networks for traffic speed prediction*. In: 2019 12th Asian control conference (ASCC). pp 1536–1541
21. Mao F, Ji M, Liu T (2016) Mining spatiotemporal patterns of urban dwellers from taxi trajectory data. Front Earth Sci 10(2):205–221. https://doi.org/10.1007/s11707-015-0525-4
22. Do LN, Vu HL, Vo BQ, Liu Z, Phung D (2019) An effective spatial-temporal attention based neural network for traffic flow prediction. Transp Res C Emerg Technol 108:12–28. https://doi.org/10.1016/j.trc.2019.09.008.
23. Yu B, Yin H, Zhu Z (2018) Spatio-temporal graph convolutional networks: a deep learning framework for traffic forecasting. In: Proceedings of the 27th international joint conference on artificial intelligence (IJCAI)
24. Zhao L, Song Y, Deng M, Li H (2018) Temporal graph convolutional network for urban traffic flow prediction method. arXiv:1811.05320
25. Zang D, Ling J, Wei Z, Tang K, Cheng J (2018) Long-term traffic speed prediction based on multiscale spatio-temporal feature learning network. IEEE Trans Intell Transp Syst PP:1–10. https://doi.org/10.1109/TITS.2018.2878068
26. Li Y, Luo J, Chow C-Y, Chan K-L, Ding Y, Zhang F (2015) Growing the charging station network for electric vehicles with trajectory data analytics. In: ICDE, pp. 1376–1387
27. Liu C, Deng K, Li C, Li J, Li Y, Luo J (2016) The optimal distribution of electric-vehicle chargers across a city. In: ICDM, pp. 261–270
28. Karamshuk D, Noulas A, Scellato S, Nicosia V, Mascolo C (2013) Geo-spotting: mining online location-based services for optimal retail store placement. In: SIGKDD, pp. 793–801
29. Zhang Z, He Q, Gao J, Ni M (2018) A deep learning approach for detecting traffic accidents from social media data. Transp Res Part C Emerg Technol 86:580–596
30. Quanjun C, Xuan S, Harutoshi Y, Ryosuke S (2016) Learning deep representation from big and heterogeneous data for traffic accident inference. In: AAAI, pp. 338–344
31. Fangzhou S, Abhishek D, Jules W (2017) Dxnat-deep neural networks for explaining non-recurring traffic congestion. In: Big Data, pp. 2141–2150
32. Lin Z, Fangce G, Rajesh K, John WP (2018) A deep learning approach for traffic incident detection in urban networks. In: ITSC, pp. 1011–1016
33. Chao C, Daqing Z, Pablo SC, Nan L, Lin S, Shijian L (2011) Real-time detection of anomalous taxi trajectories from gps traces. In: MobiQuitous, pp. 63–74

34. Jae-Gil L, Jiawei H, Xiaolei L (2008) Trajectory outlier detection: a partition-and-detect framework. In: ICDE, pp. 140–149
35. Shen M, Liu D-R, Shann S-H (2015) Outlier detection from vehicle trajectories to discover roaming events. Inf Sci 294:242–254
36. Wang Y, Qin K, Chen Y, Zhao P (2018) Detecting anomalous trajectories and behavior patterns using hierarchical clustering from taxi gps data. GEIN 7(1):25
37. Jaeyoung Jung R, Jayakrishnan, Park JY (2016) Dynamic shared-taxi dispatch algorithm with hybrid-simulated annealing. Comput-Aid Civ Infrastruct Eng 31(4):275–291
38. Ma S, Zheng Y, Wolfson O (2013) T-share: a large-scale dynamic taxi ridesharing service. In: ICDE, pp. 410–421
39. Jaw J-J, Odoni AR, Psaraftis HN, Wilson NHM (1986) A heuristic algorithm for the multi-vehicle advance request dial-a-ride problem with time windows. Transp Res Part B Methodol 20(3):243–257
40. Shuo M, Zheng Y, Ouri W (2014) Real-time city-scale taxi ridesharing. TKDE 27(7):1782–1795
41. Yan H, Favyen B, Ruoming J, Xiaoyang SW (2014) Large scale real-time ridesharing with service guarantee on road networks. VLDB, 7(14)
42. Wang S, Zhifeng Bao J, Culpepper S, Sellis T, Cong G (2017) Reverse k nearest neighbor search over trajectories. TKDE 30(4):757–771
43. Cheng P, Xin H, Chen L (2017) Utility-aware ridesharing on road networks. In: SIGMOD, pp. 1197–1210
44. Cordeau J-F (2006) A branch-and-cut algorithm for the dial-a-ride problem. Oper Res 54(3):573–586
45. Tian C, Huang Y, Liu Z, Bastani F, Jin R (2013) Noah: a dynamic ridesharing system. In: SIGMOD, pp. 985–988
46. Lee D-H, Wang H, Ruey LC, Siew HT (2004) Taxi dispatch system based on current demands and real-time traffic conditions. Transp Res Rec 1:193–200
47. Lee J, Park G-L, Kim H, Yang Y-K, Kim P, Kim S-W (2007) A telematics service system based on the linux cluster. In: ICCS, pp. 660–667
48. Zhang L, Hu T, Min Y, Wu G, Zhang J, Feng P, Gong P, Ye J (2017) A taxi order dispatch model based on combinatorial optimization. In: SIGKDD, pp. 2151–2159
49. Seow KT, Dang NH, Lee D-H (2009) A collaborative multiagent taxi-dispatch system. TASE 7(3):607–616
50. Alshamsi A, Abdallah S, Rahwan I (2009) Multiagent self-organization for a taxi dispatch system. In: ICAAMS, pp. 21–28

Chapter 9
Spatio-Temporal Supply Chains and E-Commerce

S. Vijayalakshmi, Sathya Shanmugasundaram, P. Padmanabhan, and S. Jerald Nirmal Kumar

Abstract Supply Chain Management (SCM) has emerged as a pivotal element of contemporary business strategies, deftly incorporating advancements in Machine Learning (ML) and Deep Learning (DL) to bolster market performance. In the E-commerce sector, SCM, enhanced by ML, is driving critical transformative changes, which have become particularly vital in the post-pandemic landscape. This evolution in SCM is setting new benchmarks in process efficiency, encompassing comprehensive risk mitigation and substantial reduction in operational costs. It ensures swift delivery and elevated customer satisfaction, while also offering deep insights into the automation of goods delivery within E-commerce. This is a crucial aspect in sculpting a globally competitive SCM model. By utilizing advanced ML software, supply chain managers in E-commerce can refine their portfolios and identify the most fitting suppliers, thus propelling their businesses towards greater efficiency and effectiveness. This article is dedicated to exploring these significant developments. It begins by examining the principles of spatial and temporal data analysis. On this foundation, it elaborates on the implementation of SCM through a variety of modern techniques, with a special emphasis on ML and DL applications. These techniques are instrumental in formulating a framework grounded in spatial-temporal data analysis. Conclusively, the article delineates the design and practical details of SCM, integrating diverse characteristics and the latest technological innovations.

S. Vijayalakshmi
Department of Computer Science and Engineering, Rajiv Gandhi College of Engineering, Puducherry, India

S. Shanmugasundaram
Department of Computer Applications, Saradha Gangadharan College, Puducherry, India

P. Padmanabhan
Department of Computer Science and Engineering, Greater Noida Institute of Technology, Greater Noida, Uttar Pradesh, India

S. Jerald Nirmal Kumar (✉)
Department of Computer Science and Engineering, JAIN Deemed -to- be University, Bangalore, India

J. A et al. (eds.), *Spatiotemporal Data Analytics and Modeling*, Big Data Management, https://doi.org/10.1007/978-981-99-9651-3_9

Keywords Spatio-temporal data · E-commerce · Supply chain management

9.1 Introduction

E-commerce has the ability to alleviate agricultural supply chain issues and aid in the implementation of urban and rural revival programs. A previous study has mostly concentrated on the commercial models of E-commerce platforms, with little attention paid to the processes through which they might optimize and reorganize the SCP [1]. It is a multi-functional platform that provides technical assistance, warehousing, logistics, supply chain financing, and insurance, among other services, according to the findings. It is not just a multichannel information management platform, but it also improves supply chain capabilities.

In spatio-temporal SCP, the transportation of products from one location to another is characterized by supply chain management [2]. Supply chain management is the practice of transferring material, information, and financial information among organizational divisions in order to serve customers. As a result, the entire SCP system benefits. A significant amount of time, effort, and money is spent on the logistics of raw materials needed for manufacturing, distribution of items, and return of faulty goods.

cAs previously said the fundamental limitation in supply chain management (SCM) is guaranteeing the smooth flow of commodities, whether completed products, raw materials, or returned items. The most difficult problem for transport managers is route planning and distribution network creation. Intelligent movement analysis will aid in the efficient transportation of objects. To augment the urgently needed goods, a real-time update of the location of the object in transit is required. According to surveys of renowned chief supply officers, an increasing number of businesses are investing in intelligent technologies to visualize the supply chain. Various techniques such as indexing methods and decision trees are used for managing the supply chain data [3]. The combination of SCM of data-science, predictive analytics, and big data, is identified as DPB.

9.1.1 Tracing and Mobility

Outright tracking of goods/items movement will significantly reduce product delivery interruptions, delays, and risk of damage. The exponential increase, availability, and use of location-aware technologies enable improved object tracking [4]. The rapid development of accurate context-aware technologies has made mobile system navigation exceedingly simple and precise. These technologies facilitate data-driven predictive exploration, decision-making, and smart solutions to enhance energy efficiency, resource utilization, and waste management [5]. Since the SCM system is predominantly concerned with managing containers and other carriers, the context of the moving object is a critical consideration, particularly when its delivery is both time-sensitive and reliant on location-aware systems.

9.1.2 Moving Object Big Data Management

Due to the inevitable usage of social networking in E-commerce, having a moving objects database (MOD) is highly mandatory in this field of SCM to ensure timely delivery of goods to the stakeholders. As a result of MOD, the volume of datasets has greatly increased. The data transferred to distant storage devices is in heterogeneous formats, owing to the proliferation of diverse and plentiful data. Consequently, a novel technique known as big data has emerged. Data extracted from location-based sensors associated with the data object in the supply chain includes the data object-id, location coordinates (latitude and longitude), dimensions, and the duration the object dwells in the spot. Utilizing these data, the supply chain can be visualized, and further insights may be extracted, reducing the majority of uncertainties that may arise during the shipping process [6, 7].

Businesses naturally accumulate their data across numerous storage systems, employing traditional database organization methods. Key trends such as mobile commerce and L-commerce (Location-dependent Commerce) necessitate precise tracking of moving objects' locations. Applications in these domains are required to meticulously maintain data pertaining to customers and service area locations. To support these requirements, numerous Moving Object Database (MODB) systems have been developed. These systems serve as repositories for the datasets, employing sophisticated indexing and querying techniques. These techniques are adept at handling both structured data and location-oriented information [8]. This innovative approach has garnered significant attention from both researchers and management specialists, marking a notable advancement in the field.

Queries articulated in the spatio-temporal databases enhance progressive, eloquent, and predictive analysis methods. An incessant stream of spatio-temporal data regarding the time-stamped locality of the vehicle is accumulated in an astonishing way called trajectories. The trajectory is the series of time-stamped positions and is the primary methodology for indicating moving data. The travel is symbolized in the trajectory with two spatial mechanisms and one temporal measurement.

Moving objects in SCM generates trajectory which comprises locality position and time. The travel of completed products begins for the dispatch zone of the factory, by supposing various methods of conveyance it ensues to its destination. Diverse delivery vehicles from the source can take several lanes which will have assorted speed, blockers, directions, stopping, etc. For computational persistence, the dimension of the moving object is disregarded here and each moving object in motion is preserved as points. In this framework, spatial and temporal characteristics of the moving points are significant factors for plotting the trajectory. Spatio-temporal clustering on this trajectory constructs crowds of objects that are physically and temporally adjacent [9]. The spatio-temporal moving objects data study

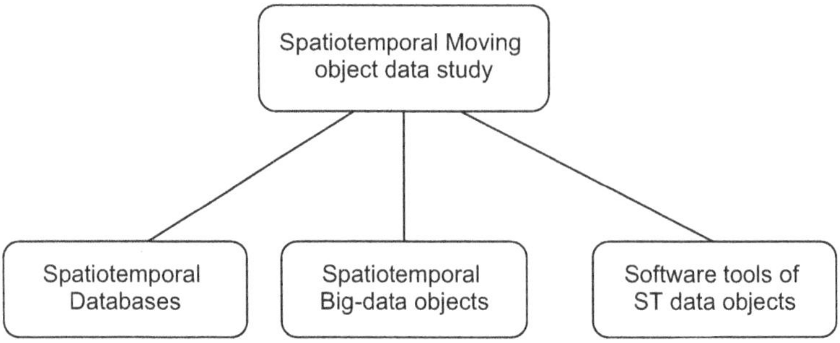

Fig. 9.1 Categorization of spatio-temporal data objects

classified into three types such as spatio-temporal database, spatio-temporal big data objects, and software tools for spatio-temporal data objects is shown in Fig. 9.1.

9.1.2.1 Spatial Moving Object (MO) Database

Spatio-temporal MO data are primarily classified into three distinct storage categories. These include: (1) Spatio-Temporal Database (ST Database) storage, (2) methodologies for handling large spatio-temporal datasets, often referred to as "ST Big Data" techniques, and (3) existing software tools designed specifically for storing and managing ST datasets. The research and development in the field of spatial and spatio-temporal database structures have evolved from traditional relational database management systems (RDBMSs) [10, 11].

Conversely, these systems tend to excel primarily with smaller datasets due to their limited capacity for handling extensive data traffic and congestion. Furthermore, these traditional schemas often fall short in terms of analytical and conceptual capabilities. As a result, spatial RDBMS (Relational Database Management Systems) prove to be less effective for storing and querying large spatio-temporal (ST) datasets, which contain substantial volumes of data. Additionally, orthodox RDBMS and No-SQL database software face limitations in storing and mining STDB (Spatio-Temporal Database) datasets. They are particularly challenged when handling unstructured data, such as that from social media platforms and sensor-generated data [12].

The conversant advantages of these widespread spatial RDBMSs are précised in Fig. 9.2. These spatial RDBMSs are developed, unwavering, and contain proficient SQL-query engines. These organizations upkeep data formats WKB and WKT cum geometry-objects like polygon, point, collections, and line string. Although most of these spatial RDBMSs support R-Tree category of indexing [11], except SQL Server and IBM DB2, where grid-indexing has been utilized.

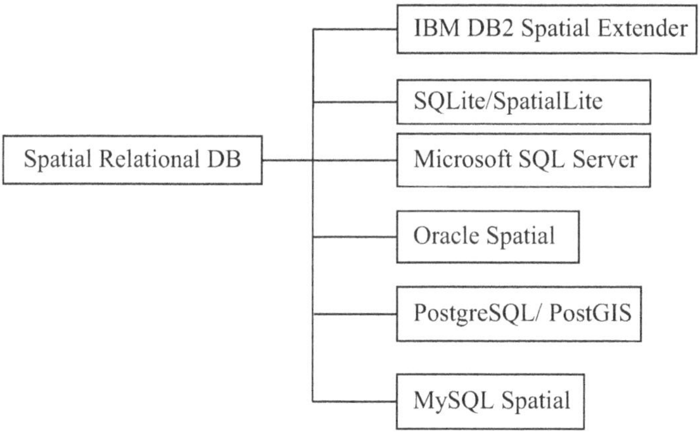

Fig. 9.2 The various tools of spatial-relations databases

9.1.2.2 Prediction of SCM's Path

Smart road traffic management systems are crucial for providing real-time traffic forecasts, especially beneficial for heavy vehicles and trucks. These systems enable drivers to find alternative routes during adverse conditions. Central to this capability is the trajectory database, which records the movement history of dynamic objects (DOs) over specified periods. Many experts propose utilizing advanced traffic prediction methods that harness spatio-temporal (ST) data collected from diverse sensors. This trajectory data, treated as a vital repository, aids in forecasting feasible routes. Intelligent systems, driven by artificial neural networks, adeptly analyze various scenarios from this authentic dataset, thus bolstering traffic management efficiency.

The utility of these systems extends to recommending optimized paths. As discussed in this section, clustering trajectory data, based on semantic attributes such as the object's route and speed, leads to more insightful analyses. Crucially, the direction of an object's movement is a key factor in clustering trajectory paths, alongside the consideration of the vehicle's speed and direction [13, 14].

9.2 Machine Learning Model for Supply Chain

By analyzing extensive historical datasets, Machine Learning (ML) algorithms can efficiently identify patterns that may be challenging or less obvious for humans to detect. This capability enables businesses to make more informed, rapid, and accurate decisions for Supply Chain Management (SCM) advancements. SCM encompasses all activities from the initial suppliers and vendors to the final satisfaction of end users.

Economic globalization and SCM have catalyzed significant changes globally in terms of finance, logistics, data recognition, and human resources. This section delves into how machine learning significantly contributes to various aspects of supply chain management. SCM typically involves five key stakeholders: customers, suppliers, products, retailers, and supervisors. The locations of these stakeholders dynamically shift based on their involvement in various activities. Spatio-temporal (ST) data objects related to these stakeholders are continuously collected and stored in STDB (Spatio-Temporal Database) servers [15, 16]. Figure 9.3 illustrates the basic structure of SCM.

The traditional SCM structure, decision-making based on dynamic data and location-based data is still one of the main issues. The finding solutions to dynamic and location-based issues of indexing and machine learning techniques are used. In this section, the machine learning, indexing method, elementary structure of SCM, and set of features are used to create an effective decision-making in SCM. The structure of the machine learning oriented decision-making is presented in Fig. 9.4. The structure of the machine learning model for SCM consists of history data and interference for removing influence, decision tree for making fruitful decision, K-NN algorithm for live data and decision-making, SARIMA for seasonable data updating, and finally forecasting and decision-making [17].

The working process of the forecasting model is performed with the help of two datasets such as History data and Live data. The working process of machine learning-based SCM model is as follows:

(i) Initially, the history data is collected and removed centered on the influences with the help of fuzzy interference. The basic decision is performed using the fuzzy interference, particularly with the help of "IF and THEN" rules or AND or OR for preliminary decision.

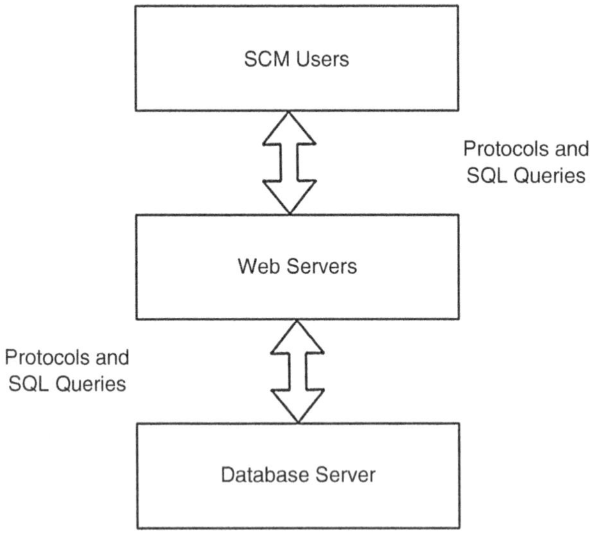

Fig. 9.3 Primary structure of SCM

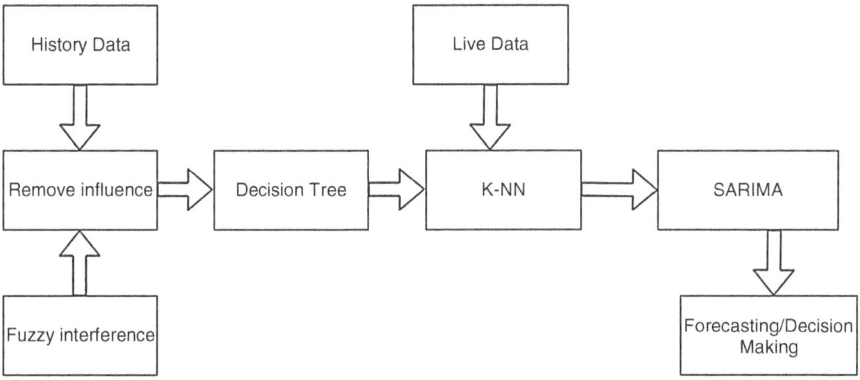

Fig. 9.4 Structure of machine learning-based SCM for forecasting/decision-making

(ii) After, the initial decision, the decision tree is applied for better decision-making. The decision tree is a supervised learning algorithm, which is used for classification and regression tasks. The decision trees divide the different strategies based on the different conditions such as spatial parameters and temporal parameters. For example, the spatio-temporal parameters are available goods in particular location, month-wise available goods in particular location and time, local location requirements, local demand, etc.

(iii) Based on the history data decision, again the live data is applied for the nearest locations information can be predicted with the help of K-NN. Using this K-NN method, the nearest information is anticipated.

(iv) Finally, the SARIMA method is applied for seasonal-based prediction. SARIMA consists of auto regression for predictions and seasonal order for forecasting of different requirements based on months, weeks, and seasonable changes. These seasonal changes goods, demand, and future information are conjectured.

Using this machine learning model for prediction and forecasting different metrics are used such as demand, frequency, tracking, real-time route, and forecasting. For forecasting and demand predictions different datasets are used such as price predictions of land, goods requirement prediction, stock prediction, and location-based requirement prediction.

9.3 Deep Learning Model for Supply Chain Management

Deep learning techniques are useful to forecast and predict the current and future requirements of information. In SCM prediction and forecasting of demand, deep learning algorithms that have been increasingly utilized in this domain due to their ability to extract patterns and insights from large and complex datasets. The different algorithms and techniques are used to analyze the SCM such as Deep Neural Network, Recurrent Neural Network, and Conventional Neural Network. The survey of [18] shows that the percentage wise contributions of the deep learning algorithms in the SCM. As per the denoted survey [18], 34% of predictions have been equipped with Conventional Neural Network in SCM. The Conventional Neural Network is used for product classification, agriculture, and supply chain predictions. Similarly, the deep neural network is instigated for analysis and predictions of quality of agriculture products, textiles requirements, and consumer requirements analysis. The recurrent neural network (RNN) is used to predict and forecast sales and financial management are different sectors.

In this section, the deep learning-based structure is presented to analyze the demand of supply chain management. Particularly, forecasting the demand of oil market using the social media presented in this example using the CNN and backpropagation neural network. The structure of the forecasting model consists of two parts such as classification of news form the media and forecasting of demand using the different parameters [19] and forecasting structure has been presented in Fig. 9.5.

The CNN's high level of accuracy makes it useful for image recognition [20]. This work is used to classify the texts from social media images. Image recognition has several uses in a variety of industries, including phones, security, recommendation systems, and medical image analysis, among others. In mathematics, convolution is the act of combining two functions to create a third function that explains how one function affects another's shape. In order to obtain information from an image, image pairs that may be described as matrix are combined to create an output. The basic convolution architecture, the feature extraction method and identifies the unique features of an image used for classification of texts from images. There are many combinations of convolutional or pooling layers in the feature extraction network. A densely integrated layer that utilizes the output of the convolutional process and constructs the image category utilizing already obtained traits. This CNN feature extraction model seeks to extract the fewest number of features from a dataset as it is practical. By integrating the qualities of an initial collection of benefits into a single new feature, it develops additional features. The three different kinds of tiers that constitute the CNN are convolution, pooling, and fully connected (FC) layers. By stacking these layers, CNN architecture is created. In addition to these three layers, there are two additional significant elements, the following is a description of the perceptron and the dropout layer.

The RNN is used to model sequential and time series data. Using this model analysis, we analyze the data like human behaviors and process the sequence data. The structure of the RNN consists of three layers such as input layers, hidden layers, and output layers. The basic architecture of the RNN consists of one-to-one, one-to-many, and many-to-one, and many-to-many architectures are used to predict data. In this work, many-to-one architecture is used to analyze the behaviors of the data [21, 22].

The working process of the architecture (Fig. 9.5) consists of two parts such as data classification and forecasting of the demand. CNNs are typically used for tasks like classification of images. From the news images, different features are collected for the classification of data. After the collection of the data, multiple other supporting data were combined. The past data and financial data are combined for future

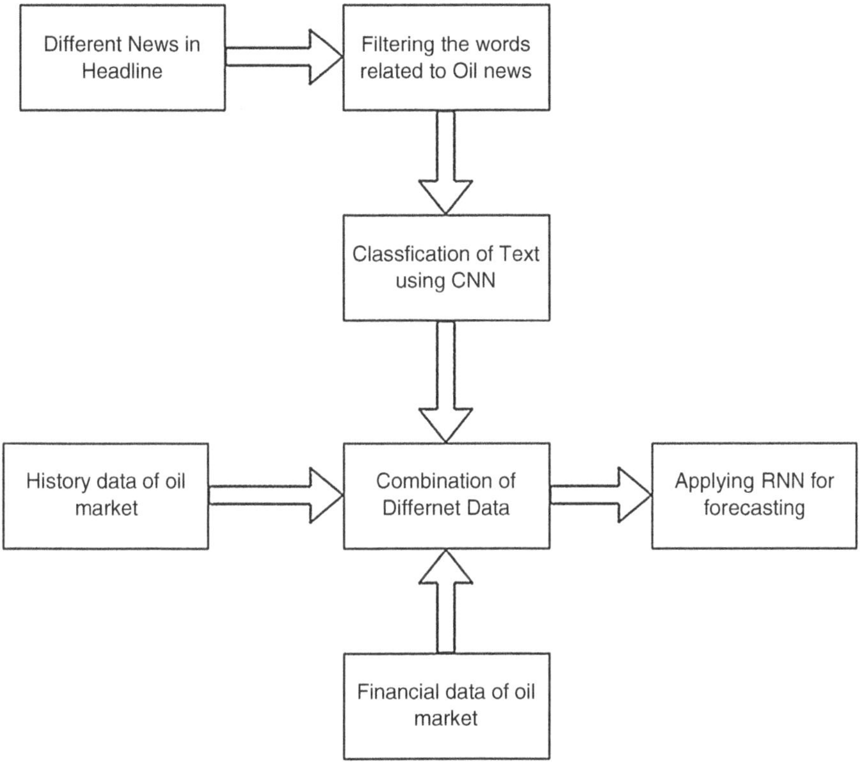

Fig. 9.5 Structure of deep learning-based SCM for forecasting

data forecasting. The RNN algorithm is applied for forecasting the demand and price of the oil marks.

9.4 General Supply Chain Management Framework

This section presents the framework for SCM using machine learning and deep learning. The SCM consists of different function such as purchasing/acquisition, operations, logistics, resource management, and information workflow [23]. Purchase and acquisition management is an important factor to fix the product cost, so proper analysis is required for getting items. Before purchasing a raw material, the spatial-temporal analysis required to get the information of raw materials production location (spatial) and temporal (timing) information. So, for analyse the purchasing information the proper spatial-temporal-based analysis is required in SCM [24, 25].

Other functions of the SCM are managing the daily basis prediction of the company materials with requirement forecasting. Before performing the purchase requirements, the proper internal requirement analysis is required based on timing. The internal requirement analysis using the timing is performed with the help of temporal analysis of data. The internal analysis requirement prediction is performed with the help of forecasting demand of the goods, availability of the raw materials in the company, and how long time supporting raw items are available. So, the operations requirement is an analysis using the temporal analysis of the data.

Logistics is an important functionality to coordinate the SCM and other functions of the production such as inventory management and warehousing of the productions. The SCM communicates and ensures the products reach the concerned location and proper timing. So, spatial analysis such as traffic prediction, travel estimation time to a particular location, and other factors with temporal features are important to manage the logistic analysis of SCM. In the operation of the SCM, three important factors work together to complete the operations such as logistics, suppliers, and intermediates. So these factors influenced the SCM has effectively worked.

The other factor affecting the SCM is resource management which consists of organizing, planning, and controlling of resources. Resource management includes raw materials from the spatial location, labor, and technology to finish the products. In resource management, reducing the requirement of consumption is an important task. So, an effective optimization is required to optimize the entire SCM. The different activities need to consider in the optimization process such as outsource reduce the communication between suppliers and retailers, usage of effective technology, multichannel approach to demand supply analysis, future forecast, customer demand forecast, etc.

Another process of SCM is workflow management. The workflow management in the SCM is related to how and what kind of information flow or communication is there between the SCM. If the communication is not effective, the result of disruptions, mistakes, and delay will occur. So, location based and time based effective workflow is required using the different factors within right time and place. For example [26], GIS spatial analysis with spatial-temporal technology help to determine the spatial distribution of tea trade of location-based enterprise supply chain management.

Different machine learning and deep learning techniques are used to analyze and forecast the spatial-temporal data using different spatial-temporal features. The general structure of SCM-based machine and deep learning are presented in Fig. 9.6. Some of the machine learning and deep learning techniques of analysis of spatial-temporal are as follows:

In machine learning, supervised and unsupervised algorithms are used to forecast and analyze the spatial-temporal data in SCM [27, 28]. The supervised algorithms is based on the labeled data, so this method incorporates a training set of patterns related to label data, learning and predicting are made simple for the algorithm from input to output. Examples of supervised learning include classification methods like Naive Bayes, SVM, linear, K-NN, Decision tree, as well as Random Forest used information from both the input and the outcome to make a forecast. Unsupervised learning is often referred to as grouping of similar data in the spatial

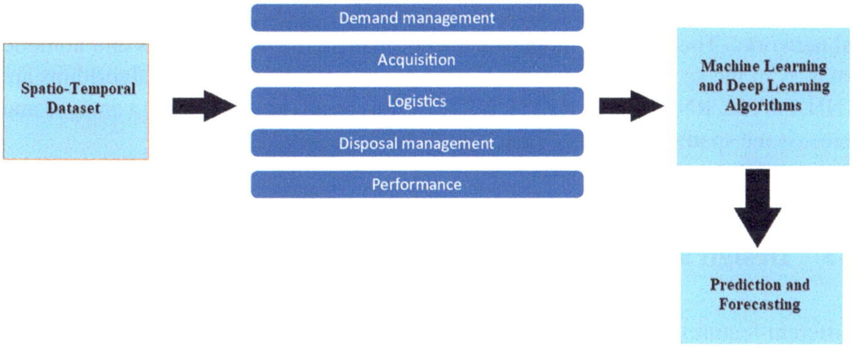

Fig. 9.6 General framework for SCM

location. In unregulated instruction, there is no label, no dataset for training, and no known data for output. The guidance is favoring an independent learning strategy and observed by utilizing clustering methods such as PCA and K-Means clustering SVD. This kind of machine learning is agent and action-based. Devices and algorithms that automatically define particular behaviors based on environment, motivation, and condition are determined by the incentive feedback in the context.

SVM is a well-versed technique for both regression and classification. The calculation generates an ideal hyperplane using the meticulously collected data that arranges fresh samples. In such case, SVM classification performed better than Naive Bayes classification. A non-parametric classification method is K-Nearest Neighbor. Data is classified by determining which of the k examples in the K-NN algorithm resembles an input the most. A measurement is used to determine this distance. A method called a decision tree includes specifically, criteria concerning dataset, could be "true" or "false" in the spatial-temporal dataset. To predict the class at the leaf, rules can be obtained using a path that travels from root to the leaf while also using the nodes in motion as prerequisites. The random forest-based algorithms come under supervised algorithms and they work based on the set of training dataset with specified features. Both classification and regression may very well benefit from its use. A forest with many trees is essentially created by a random forest tree. It is therefore essentially a collection of various decision trees. The classification process uses a vote system to determine the class. The combination of different learning algorithms has created better model or algorithms for better predictions. Ensemble classification techniques are frequently utilized in the therapeutic sector to achieve greater accuracy than using a unique categorization. It is, essentially, a machine learning technique that enhances predictive performance by combining several machine learning algorithms.

A neural network with some level of difficulty, typically at least one or two layers, is characterized as a deep neural network (DNN), or deep net. Deep neural networks process information [18] in various ways by using sophisticated mathematical models. In order to address practical problems like categorization, a neural network's basic purpose is to take in a variety of inputs, process them using more difficult calculations, and output the findings. We are restricted to feed-forward neural networks. The different models are used to analyze image datasets and numeric datasets in the spatial-temporal datasets. Some of the deep learning algorithms are CNN, LSTM, RNN, GAN, RBFN, MCP, SOMs, and DBNs are used to analyze and forecast the spatial-temporal datasets in SCM.

9.5 Design and Implementation of SCM System

Different features, functional activities, and technical insights are taken into consideration in the design and implementation of SCM. Enterprise management is processed using a variety of modern technologies with the features, functional activities, and technical activities.

Some of the characteristics of the SCM consists of expansibility, modularity, openness, ease of use, and collaborations. Expansibility implies that the system must be able to accommodate new requirements and modify itself to meet related business needs. And also, system should adapt for user new fields and extend the secondary development. Modularity means each subsystem should be independent, without affecting other modules of the SCM system. Multiple subsystems should share the information throughout the whole process and data integration of SCM. Openness means the integration of information with different data exchange in each module openly for continuous access of each node using different software. Ease of use means the system should be easy to deployment without contactless, on-demand automatic installation of requirements, easy access to client-server software, etc.

Different recent technologies are used to implement the functionalities of SCM. The functionalities are implemented in two ways such as vendors level and administration level. From the administrative level to the client level, all data are updated and merged by the vendors. The accuracy, speed, quality, price, etc. are measured with the help of recent software techniques. A variety of administrative level tasks are kept an eye on, including automated product checking, condition monitoring, prediction analysis, future forecasting of demanding, and managing warehousing are monitored. And also, the effective data interface is used to manage the data warehouse based on the spatial-temporal data with the help of GPS and GIS systems. Three tire architectures are utilized in the technical realizations of the program, database server, and browser in the user and also, implementation is performed in the level of LAN and Wide area network. The approach with spatial-temporal aspects essentially uses a client-server structure that used in the implementation with spatial-temporal features with Machine and Deep Learning techniques are applied in framework and application implementations.

9.6 Summary

This chapter covered the spatial-temporal integration of supply chain management in e-commerce environments. The fundamental information and essential requirements of supply chain management using spatial-temporal data are presented in the entire chapter. The basic tracing and mobility of the spatial-temporal are given and also moving object big data management in the SCM exists. The supply chain management in the prediction of logistics along with trajectory is done by using spatio-temporal data objects. The functional requirements, machine and deep learning techniques in the SPM with different implementation characteristics are presented in this chapter.

Deep learning-oriented structure to analyze the demand of supply chain management is also incorporated in this chapter. Predominantly, forecasting the demand of current market strategy using the CNN and backpropagation neural network. The general framework, design, and implementation of SCM and its functionalities integrated with machine learning are also summarized.

References

1. Al Mashalah, Heider, et al. "The impact of digital transformation on supply chains through e-commerce: Literature review and a conceptual framework." Transportation Research Part E: Logistics and Transportation Review 165 (2022): 102837.
2. Krishnaveni, K. S., and P. P. Anil Kumar. "Spatio-temporal dynamics of urban sprawl in a rapidly urbanizing city using machine learning classification." Geocarto International 37.27 (2022): 17403-17434.
3. John, A., M. Sugumaran, and R. S. Rajesh. "Indexing and query processing techniques in spatio-temporal data." ICTACT Journal on Soft Computing 6.3 (2016): 1198-1271.
4. Car, Tomislav, Ljubica Pilepić, and Mislav Šimunić. "Mobile technologies and supply chain management-lessons for the hospitality industry." Tourism and hospitality management 20.2 (2014): 207-219.
5. Jagtap, Sandeep, et al. "IoT technologies in the food supply chain." Food technology disruptions. Academic Press, 2021. 175-211.
6. Anderson-Grégoire, Isabelle M., et al. "A big data science solution for analytics on moving objects." International Conference on Advanced Information Networking and Applications. Cham: Springer International Publishing, 2021.
7. A. John, M. Sugumaran and R. S. Rajesh, "Performance analysis of the past, present and future indexing methods for spatio-temporal data," 2017 2nd International Conference on Communication and Electronics Systems (ICCES), Coimbatore, India, 2017, pp. 645-649, doi: https://doi.org/10.1109/CESYS.2017.8321157.
8. Laksmiwati, Hira, et al. "Modeling unpredictable data and moving object in disaster management information system based on spatio-temporal data model." 2014 International Conference on Data and Software Engineering (ICODSE). IEEE, 2014.
9. Atluri, Gowtham, Anuj Karpatne, and Vipin Kumar. "Spatio-temporal data mining: A survey of problems and methods." ACM Computing Surveys (CSUR) 51.4 (2018): 1-41.
10. Koperski, Krzysztof, Jiawei Han, and Nebojsa Stefanovic. "An efficient two-step method for classification of spatial data." proceedings of International Symposium on Spatial Data Handling (SDH'98). 1998.

11. I. Sasikala, M. Ganesan and A. John, "Uncertain data prediction on dynamic road network," International Conference on Information Communication and Embedded Systems (ICICES2014), Chennai, India, 2014, pp. 1-4, doi: https://doi.org/10.1109/ICICES.2014.7033972.
12. Md Mahbub Alam, Luis Torgo, Albert Bifet. "A Survey on Spatio-temporal Data Analytics Systems", ACM Computing Surveys, 2022.
13. Malviya, Rakesh Kumar, and Ravi Kant. "Hybrid decision making approach to predict and measure the success possibility of green supply chain management implementation." Journal of Cleaner Production 135 (2016): 387-409.
14. John, A., et al. "Dynamic sorting and average skyline method for query processing in spatial-temporal data." International Journal of Data Science 6.1 (2021): 1-18.
15. Ghazal, T. M., and H. M. Alzoubi. "Modelling supply chain information collaboration empowered with machine learning technique." Intelligent Automation & Soft Computing 29.3 (2021): 243-257.
16. Lin, Haifeng, Ji Lin, and Fang Wang. "An innovative machine learning model for supply chain management." Journal of Innovation & Knowledge 7.4 (2022): 100276.
17. Mohan, Senthilkumar, et al. "An approach to forecast impact of Covid-19 using supervised machine learning model." Software: Practice and Experience 52.4 (2022): 824-840.
18. Hosseinnia Shavaki, Fahimeh, and Ali Ebrahimi Ghahnavieh. "Applications of deep learning into supply chain management: a systematic literature review and a framework for future research." Artificial Intelligence Review 56.5 (2023): 4447-4489.
19. Wu, Binrong, et al. "Forecasting the US oil markets based on social media information during the COVID-19 pandemic." Energy 226 (2021): 120403.
20. Makantasis, Konstantinos, et al. "Deep supervised learning for hyperspectral data classification through convolutional neural networks." 2015 IEEE international geoscience and remote sensing symposium (IGARSS). IEEE, 2015.
21. Medsker, Larry R., and L. C. Jain. "Recurrent neural networks." Design and Applications 5.64–67 (2001): 2.
22. Tenti, Paolo. "Forecasting foreign exchange rates using recurrent neural networks." Applied Artificial Intelligence 10.6 (1996): 567–582.
23. Ho, Danny CK, K. F. Au, and Edward Newton. "Empirical research on supply chain management: a critical review and recommendations." International journal of production research 40.17 (2002): 4415–4430.
24. Soni, Gunjan, and Rambabu Kodali. "A critical review of supply chain management frameworks: proposed framework." Benchmarking: an international journal 20.2 (2013): 263–298.
25. Liu, Shaofeng, et al. "A knowledge chain management framework to support integrated decisions in global supply chains." Production Planning & Control 25.8 (2014): 639–649.
26. Zhang, Shiliang, and Tingcheng Chang. "Spatial–temporal evolution of the distribution pattern of customer sources in tea trade of Fujian enterprise supply chain." Microsystem Technologies 27 (2021): 1305–1315.
27. Yang, Mei, et al. "Supply chain risk management with machine learning technology: A literature review and future research directions." Computers & Industrial Engineering (2022): 108859.
28. Park, Kyoung Jong. "Determining the tiers of a supply chain using machine learning algorithms." Symmetry 13.10 (2021): 1934.

Chapter 10
Spatiotemporal Renewable Energy Techniques and Applications

Abhishek Vyas, Satheesh Abimannan, Po-Ching Lin, and Ren-Hung Hwang

Abstract This chapter provides an overview of the field of spatiotemporal data analysis in the context of renewable energy applications. With the increasing use of renewable energy resources, more advanced and accurate analysis of spatiotemporal data is needed to optimize energy systems. The chapter discusses the importance, limitations, and applications of spatiotemporal data analytics, including predicting solar and wind energy production, analyzing the impact of weather events on renewable energy systems, and optimizing the placement and performance of renewable energy systems. One of the biggest challenges, however, is the volume and complexity of the data, which includes weather patterns, energy production, and use. Spatiotemporal data analytics for renewable energy applications is rapidly growing, but little is known about its potential benefits and limitations. More research is needed to fully understand how this technology can improve the reliability, efficiency, and sustainability of renewable energy systems. The chapter provides readers with a comprehensive view of the methodologies, algorithms, datasets, techniques, and frameworks used by various researchers in their state-of-the-art work on the applications of spatiotemporal data analytics in renewable energy systems. This chapter also discusses future research directions to improve the availability of renewable energy systems worldwide.

Keywords Renewable · Energy · Data analytics · Spatiotemporal data

A. Vyas (✉) · P.-C. Lin
Computer Science and Information Engineering Department, National Chung Cheng University, Chiayi County, Taiwan
e-mail: abhishek_vyas_2022_7@ieee.org; d09410004@ccu.edu.tw

S. Abimannan
Amity School of Engineering and Technology, Amity University Maharashtra, Mumbai, Maharashtra, India

R.-H. Hwang
College of Artificial Intelligence, National Yang Ming Chiao Tung University, Tainan, Taiwan

J. A et al. (eds.), *Spatiotemporal Data Analytics and Modeling*, Big Data Management, https://doi.org/10.1007/978-981-99-9651-3_10

10.1 Introduction

Alternative or renewable energy resources, like wind energy, solar energy, hydrological energy, and geothermal energy, have become increasingly important as alternatives to traditional forms of energy. However, their intermittent nature and geographical variability pose significant challenges to their integration into the power grid. Spatial-temporal data analysis can help overcome these challenges by providing insights into the availability and variability of renewable energy resources. This approach involves the use of advanced analytical techniques to analyze and interpret data collected from various sources, including weather stations, satellite imagery, and sensor networks. By leveraging this data, renewable energy stakeholders can optimize their operations, increase efficiency, and make more informed decisions about resource allocation and infrastructure development. In this chapter, we will explore the various applications of spatial-temporal data analysis in renewable energy and how it is transforming the way we generate and consume energy.

Alternative energy resources have become a promising solution to the world's increasing energy demands while lowering the harmful effects of fossil fuel use on the changing global climate. However, alternative energy resources are often located in remote and geographically diverse areas, and their power generation varies based on weather patterns, seasons, and other factors. Therefore, it is challenging to integrate renewable energy into the grid, and stakeholders must analyze vast amounts of spatial-temporal data to overcome these challenges.

Spatial-temporal data analysis leverages advanced techniques, including machine learning algorithms and statistical models, to extract insights from geographically and temporally dispersed datasets. For instance, weather stations collect a wealth of data on temperature, humidity, wind speed, and solar irradiance, among others, essential variables for renewable energy production. By integrating this data into predictive models, stakeholders can forecast the production of alternative energy resources and optimize their utilization accordingly.

Satellite imagery is another crucial data source for renewable energy applications. It can provide valuable information on land cover, land use, and topography, enabling stakeholders to identify conducive locations for alternative energy projects, such as wind turbine farms, solar panel farms, and hydroelectric power plants. Sensor networks are useful for capturing raw data on energy production, transmission, and usage, allowing stakeholders to monitor their operations and make informed decisions.

The utilization of spatial-temporal data analysis holds significant potential in enhancing the efficiency and performance of alternative energy systems, mitigating operational costs, and bolstering the reliability and stability of the power grid. For instance, through predictive models, it becomes possible to effectively manage the variable nature of renewable energy resources by accurately forecasting energy generation and anticipating fluctuations in demand. Furthermore, stakeholders can employ spatial-temporal data analysis to continually monitor the performance of renewable energy systems, identifying areas that require enhancement. The applications of this analytical approach in renewable energy span a broad spectrum, encompassing diverse aspects. By capitalizing on this methodology, stakeholders can deepen their comprehension of renewable energy resources, optimize their operations, and actively contribute towards a more sustainable energy future.

Now to come to terms with the concept of spatiotemporal data analysis, let us delve into the significant terms and concepts of spatiotemporal data analytics, as outlined in [1]. At present, the landscape of spatial and spatiotemporal data analytics systems can be classified into three primary categories: (1) spatial databases, (2) infrastructure for processing large-scale spatial data, and (3) programming languages and software for geographic information systems [1]. Figure 10.1 shows the ecosystem for the spatiotemporal data analytics in a graphical format.

Now let us derive a definition of spatiotemporal data from basics. "Raster data is represented as a collection of pixels associated with a specific geographical location. Raster data can be discrete or continuous. Points, lines, or polygons can represent vector data. A spatial network is a special graph that consists of nodes embedded in space. The most common example of a spatial network is the transportation network, where edges represent road segments, and nodes represent the intersection of road segments or points of interest [1]."

However, data with spatial features being captured with timestamps is called spatiotemporal data. Spatiotemporal data consists of an object's spatial and temporal features. Spatiotemporal data type, in general, is nothing but an amalgamation of timestamps with the spatial data type. Now, technically, the spatial data is stored in two types of databases: (1) relational SQL databases and (2) non-SQL databases; it is worth noting here that SQL means structured query language, and non-SQL databases include graph databases, document-based databases, key value-based databases, and column family databases. Some popular SQL databases are MySQL

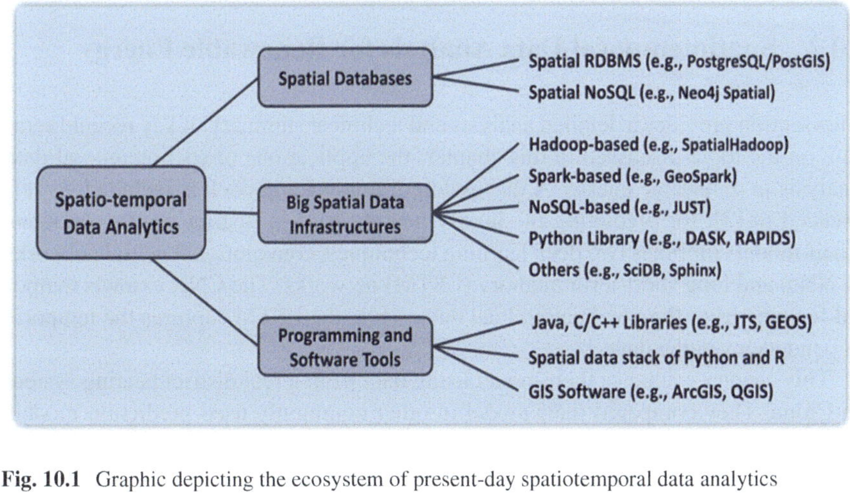

Fig. 10.1 Graphic depicting the ecosystem of present-day spatiotemporal data analytics

Spatial, MS SQL Server, IBM DB-2 Spatial Extender, PostGIS, Oracle-Spatial, and SQLite-based SQL-Spatial [1].

Some of the basic and required programming and software tools for spatiotemporal data analytics are the following: (1) the R-programming language, its allied packages, and APIs; (2) the Python-programming language, its allied packages, and APIs; (3) GIS software packages like ArcGIS, GRASS, QGIS, and SAGA; and finally (4) Java-based Hadoop cluster that is also very helpful in spatiotemporal data analysis.

In addition to the aforementioned software and database tools, spatiotemporal data analytics in diverse domains, particularly in the renewable energy field discussed in this chapter, extensively leverage machine learning (ML) and deep learning (DL) algorithms and techniques [1].

In the preceding sections, we introduced the renewable energy domain and its use cases where the application of spatiotemporal data analytics is helpful and sometimes deemed necessary. Additionally, we provided readers with a comprehensive background on spatiotemporal data, including its definitions, meanings in various contexts, and the diverse tools and techniques used for its analysis.

In the upcoming sections, we present a well-organized scientific review of the state-of-the-art research conducted recently in this field. This review focuses specifically on the application of spatiotemporal data analysis in addressing issues related to renewable energy.

This chapter is structured as follows. Section 10.2 discusses the peer-reviewed literature published in recent years on the topic at hand. Section 10.3 presents a comprehensive discussion of the methods and technologies seasoned researchers use in spatiotemporal analytics for renewable energy applications. Section 10.4 is where we discuss our views based on the literature survey and outline future research directions. This section specifically focuses on the research concerning the use of spatiotemporal data analytics in applications for renewable energy resources.

10.2 Spatiotemporal Data Analysis for Renewable Energy

This section provides a detailed analysis and technical summary of key recent literature on the topic discussed in this chapter: the applications of spatiotemporal data analysis in renewable energy. A discussion of a novel approach is included, as referenced in [2], for predicting the hourly heating load in district heating systems. Their model combines two deep learning techniques, convolutional neural networks (CNNs) and long short-term memory (LSTM) networks. The CNN extracts temporal features from the raw heating load data, while the LSTM captures the temporal dependencies in the data.

This chapter evaluates their model using data from a real district heating system in China. They compared their model to other commonly used prediction models and found that their hybrid CNN-LSTM model outperformed the others in terms of accuracy and stability. This makes it a promising tool for predicting heating loads in district heating systems.

In [2], the main contributions of the chapter are outlined as follows:

- They proposed a new forecasting model based on hybrid CNN-LSTM for the heating load.
- The proposed model integrates the merits of the parallel processing ability of the CNN and the feature extraction ability of the LSTM for time series data.
- The CNN model used in the proposed model can extract nonlinear features of the heating load more accurately in the spatial domain.
- The LSTM algorithm used in the proposed model is more accurate in modeling the changing trend of heating loads in the temporal domain.
- The proposed hybrid CNN-LSTM algorithm exhibits very good adaptability to the heating load data from industrial applications.
- A comprehensive comparison of parameter optimization using the CNN-LSTM algorithm was conducted against several benchmark models.
- The CNN-LSTM model exhibits higher prediction performance compared with several traditional conventional models.

Additionally, in [3], the generation of spatiotemporal scenarios for alternative energy generation is explored. This involves preserving the privacy of participating parties by applying a federated deep generative approach for learning. The proposed method consists of two phases: (1) a centralized training phase and (2) a decentralized generation phase. In the centralized training phase, a centralized server trains a deep generative model using a large dataset of alternative energy generation data. A deep generative model is designed to generate spatiotemporal scenarios of alternative energy generation, given a set of input parameters such as weather conditions, time of day, and geographic location.

In the decentralized generation phase, a set of participating parties, each with their own dataset of renewable energy generation data, use their local datasets to generate scenarios using the trained generative model without revealing any private data to the other parties or the centralized server. This is achieved through a federated learning approach, where the generative model is trained locally on each party's data and then combined to generate scenarios without exchanging any raw data. To further ensure privacy, differential privacy techniques are employed during training. These techniques involve adding noise to the gradients used for updating the generative model, thereby ensuring that the training process does not reveal sensitive information about the individual datasets.

The proposed method is evaluated on real-world renewable energy generation datasets from three different regions in China, and the results show that it can generate accurate spatiotemporal scenarios while preserving the privacy of the participating parties. This novel approach is significant for generating spatiotemporal scenarios for renewable energy generation, addressing privacy concerns in a federated learning setting. The authors believe that the proposed method holds considerable potential, especially for other spatiotemporal forecasting tasks where privacy is paramount [3].

Moreover, an overview [4] of wind and wave energy as potential renewable energy sources is provided. This includes a discussion of the limitations of traditional modeling approaches for these energy sources and introduces the potential of deep learning techniques to overcome these challenges. The literature on applying deep learning in wind and wave energy is comprehensively reviewed, covering a wide range of topics. These include wind and wave forecasting, power output prediction, condition monitoring, fault diagnosis, and control optimization.

The section on wind and wave forecasting discusses the use of RNNs and CNNs for short-term and long-term wind and wave forecasting. It also discusses the use of hybrid models that combine traditional modeling approaches with deep learning techniques. The section on power output prediction describes the use of deep learning techniques for predicting the power output of wind and wave energy systems. It also discusses the use of feedforward neural networks (FNNs), RNNs, and CNNs for this task and also covers deep learning for wind turbine load prediction. The section on condition monitoring and fault diagnosis discusses the use of deep learning for detecting and diagnosing faults in wind and wave energy systems. It also covers this task's use of autoencoders, deep belief networks (DBNs), and convolutional autoencoders (CAEs). Finally, the section on control optimization discusses the use of deep reinforcement learning (DRL) for optimizing the control of wind and wave energy systems. It also covers the use of DRL for wind turbine pitch control, wave energy converter control, and wind farm layout optimization.

Overall, this provides a comprehensive review of the existing literature on applying deep learning in wind and wave energy. It addresses the limitations of traditional modeling approaches and proposes a road map for future research in this domain [4]. An overview [5] is also provided on various forecasting models that can be utilized for predicting electricity demand and renewable energy production in smart grids and smart buildings. This overview evaluates different techniques, their advantages and disadvantages, and their applications within the context of smart grids and smart buildings.

The chapter is structured in five main sections. In the first section, the importance of forecasting in smart grids and buildings is discussed, highlighting the challenges that arise due to the variability of renewable energy sources and the need for accurate demand prediction. In the second section, the different types of forecasting models are presented, including time series models, artificial neural networks (ANNs), fuzzy logic, and hybrid models that combine different techniques. In the third section, several methods that focus on forecasting electricity demand in buildings are reviewed, using different data sources such as historical energy consumption, weather data, and building characteristics. It also discusses the advantages and disadvantages of each technique and highlights the need for accurate data collection and processing. In the fourth section, focus is given on renewable energy forecasting, discussing the different techniques used to predict the output of solar, wind, and hydroelectric power plants. It also reviews different techniques that propose methods to optimize the integration of renewable energy sources into the grid, considering factors such as power storage and load balancing. Finally, in the fifth section, the challenges and future directions for forecasting in smart grids and buildings are discussed, emphasizing the need for accurate models that can account for the variability of renewable energy sources and provide real-time predictions for decision-making purposes. Overall, in this chapter, a comprehensive overview of the different techniques and models that can be used for forecasting in the smart grid and smart building context is provided. This chapter highlights the importance of accurate data collection and processing and the need for models that can account for the variability of renewable energy sources [5].

The systematic and organized [6] review of recent applications of machine learning (ML) in sustainable energy systems (SESs) and the challenges that must be overcome to improve their efficiency and effectiveness are discussed in this chapter. The chapter also provides a comprehensive view of the latest research in the field of ML applied to various aspects of SES, including energy generation, distribution, storage, and consumption. It also highlights the potential benefits of using ML in SES, including improved forecasting, optimization, fault detection, and energy management. The various types of ML algorithms, such as neural networks (NNs), decision trees (DTs), support vector machines (SVMs), and clustering techniques, are explained, along with examples of their use in SES. The chapter also identifies several challenges that must be addressed to fully realize the potential of ML in SES, including data availability, data quality, computational complexity, model interpretability, and system scalability. The solutions to these challenges are also provided, such as data fusion, feature engineering, model selection, and explainable AI. In a nutshell, this chapter provides a valuable and insightful overview of the current state of the art in ML applications in SES and highlights the potential for further research and development in this area [6].

The chapter studies the feasible space [7] of a renewable power system model. The method for characterizing the near-optimal feasible space of such a model is discussed, which provides valuable insights into the behavior of the system under different conditions. The renewable power system model includes variables such as power output from renewable sources, energy storage levels, and demand. They then use a mixed-integer linear programming (MILP) approach to optimize the system, subject to various constraints such as energy balance and capacity limits. By solving this optimization problem for a range of input parameters, they can generate a set of optimal solutions, each representing a unique combination of input parameters that satisfies all constraints. Next, the concept of the feasible space is introduced, which is the set of all input parameter combinations that satisfy the constraints of the model. They use a technique called principal component analysis (PCA) to identify the most significant factors affecting the feasible space. Then, they use a clustering algorithm to group the optimal solutions into clusters based on these factors. By analyzing the distribution of these clusters within the feasible space, insights into the system's behavior under different conditions are gained. Finally, the concept of the near-optimal feasible space is introduced, which includes all input parameter combinations within a certain distance of an optimal solution. By characterizing the near-optimal feasible space, regions of the input parameter space can be identified, where the system is most sensitive to changes and can thus focus their efforts on improving the system's performance in these regions. Overall, this chapter presents a valuable contribution to the field of renewable power system modeling by introducing a novel approach for characterizing the feasible space of such models and providing insights into the behavior of the system under different conditions.

A probabilistic [8] wind power forecasting model is developed for very-short-term periods (up to half an hour ahead) using sparse machine learning and nonparametric density estimation algorithms. The model is designed to estimate the uncertainty of wind power forecasts, which is crucial for power system operators to manage power system stability and ensure reliable electricity supply. First, develop a sparse regression model using the least absolute shrinkage and selection operator (LASSO) algorithm, which is a machine learning technique that performs both variable selection and regularization to prevent overfitting. The LASSO algorithm is applied to select the most relevant predictors for the wind power forecast, which are then used to build the regression model. To estimate the uncertainty of the wind power forecast, the nonparametric density estimation algorithm based on the kernel density estimation (KDE) technique is used. KDE is a nonparametric method that estimates the probability density function of a random variable based on a sample of observations. The KDE algorithm is used to estimate the probability density function of the error term in the regression model, which is then used to compute probabilistic wind power forecasts. The proposed model is evaluated using real wind power data from a wind farm located in Sweden. The results show that the proposed model outperforms several benchmark models in terms of forecast accuracy and uncertainty estimation. The model is also shown to be computationally efficient, making it suitable for real-time applications. In a nutshell, the chapter presents a novel approach to probabilistic wind power forecasting using sparse machine learning and nonparametric density estimation algorithms. The proposed model is shown to be effective in estimating the uncertainty of wind power forecasts, which is essential for power system operators to ensure a reliable electricity supply.

The deep learning-based [9] hybrid approach for solar irradiance forecasting combines two popular deep learning models: long short-term memory (LSTM) and convolutional neural network (CNN). The proposed model aims to improve the accuracy and reliability of solar irradiance forecasting, which is crucial for the effective integration of solar energy into the power grid. The proposed model comprises three main components: data preprocessing, feature extraction, and model training. In the data preprocessing stage, the raw solar irradiance data is preprocessed and normalized to remove any outliers and ensure consistency in the data. Then, in the feature extraction stage, the preprocessed data is transformed into meaningful features using a combination of time-series analysis techniques and image-processing techniques. The extracted features are then fed into the LSTM-CNN hybrid model for training and prediction. The LSTM-CNN hybrid model is designed to capture both the temporal and spatial patterns in the solar irradiance data. The LSTM component is responsible for modeling the temporal dependencies in the data, while the CNN component is responsible for modeling the spatial correlations between the different regions of the solar irradiance data. The two components are combined in a hybrid architecture that allows them to work together to produce accurate and reliable predictions. The proposed model was evaluated using real-world solar irradiance data collected from four locations in China. The results showed that the LSTM-CNN hybrid model outperformed several other state-of-the-art models, including traditional time-series models and other deep learning models, in terms of forecasting accuracy and reliability. The proposed model achieved a mean absolute percentage error (MAPE) of less than 10%, which is considered highly accurate for solar irradiance forecasting. Thus, a novel and effective approach for solar irradiance forecasting is required that combines the strengths of LSTM and CNN models in a hybrid architecture. The proposed model has the potential to improve the integration of solar energy into the power grid and advance the development of sustainable energy systems.

A comprehensive review [10] of big data analytics (BDA) techniques is given that can be applied in future electricity grids. An overview of the current state of electricity grids is also provided, including their challenges and opportunities. The chapter then discusses the potential benefits of BDA in addressing these challenges, such as improving grid reliability, enhancing energy efficiency, and reducing costs. The various BDA techniques are discussed, including data preprocessing, data mining, machine learning, and deep learning. The challenges associated with applying BDA techniques in the context of electricity grids are also explained, such as data quality, data privacy, and computational complexity. Furthermore, examples of real-world applications of BDA in electricity grids are provided, including load forecasting, fault diagnosis, and demand response. Overall, a valuable overview of the potential applications of BDA in electricity grids is given, highlighting the importance of continued research in this area.

The use of cyber-physical systems (CPSs) [11] improves building energy management. Specifically, the use of digital twin technology and artificial intelligence (AI) algorithms helps to create a more efficient and effective building energy management system. The authors begin by discussing the challenges of traditional building energy management systems, which often rely on manual data collection and analysis. They argue that CPS, which combines physical and computational components, can provide more accurate and real-time data for energy management. The chapter then describes the concept of a digital twin, which is a virtual replica of a physical system. The digital twin of a building is used to collect and analyze data in real time, using sensors and other data sources. To make use of this data, AI algorithms are used to optimize building energy usage. They suggest using machine learning algorithms to analyze data from the digital twin and make predictions about future energy usage. This would allow building managers to make adjustments in real time to optimize energy usage and reduce costs. Finally, the chapter discusses the potential benefits of using CPS for building energy management, including improved energy efficiency, reduced energy costs, and better building performance overall.

Now to change track for a bit, let us talk about the financing of renewable energy projects by various nations around the world and its technical analysis. This is done in [12] by trying to explore the financing patterns of clean energy projects and the factors that affect such financing in major investment countries, namely the USA, China, Japan, and Germany. The collected data revealed that clean energy projects are financed through various sources such as public funds, private equity, venture capital, and debt financing, among others. The study used regression analysis to examine the relationship between clean energy financing and various factors such as the level of economic development, policy support, financial market development, and environmental factors. The panel dataset covered the period from 2005 to 2018 and included both country-level and project-level variables. The results of the study showed that the levels of economic development, financial market development, and policy support were significant determinants of clean energy financing in all the countries studied. Additionally, it was found that the type of financing varied across countries, with the USA and China relying more on public funds and venture capital, while Japan and Germany favored debt financing. And it was also found that environmental factors, such as the availability of renewable energy resources, were not significant determinants of clean energy financing, suggesting that policy support and financial market development were more important in driving clean energy investments.

Above, we have identified some major research literature on the topic at hand and tried to summarize them in a reader-friendly and comprehensible manner. Now for the rest of the literature, we organize the salient features of the literature in a tabular format. This will give the readers more insight into the topic of the use of spatiotemporal data analytics in the context of renewable energy by various researchers around the world. Table 10.1 is shown here.

10.3 Techniques for Renewable Energy Applications

In this section, we will discuss some of the important methods and technologies used by the researchers in their state-of-the-art work on the use of spatiotemporal data analytics for renewable energy applications. The few important techniques and methods as the others are extracted from these techniques or are derived from them using some statistical or mathematical methodologies. Below, we discuss several important methods and technologies commonly used by researchers working on spatiotemporal data analytics for applications in renewable energy.

1. *Long Short-Term Memory (LSTM)*: LSTMs are recurrent neural networks (RNNs) specifically designed to address the vanishing gradient problem in standard RNNs. The vanishing gradient problem arises when the gradients become too small during backpropagation, which makes it difficult to learn long-term dependencies in sequential data. LSTM networks use a set of specialized memory cells to store information over time and a set of gating mechanisms to control the flow of information in and out of the memory cells. The memory cells can selectively retain or forget information based on the input data, and the gating mechanisms allow the network to learn how to selectively update the memory

Table 10.1 Tabular summary of the salient features of the literature available on spatiotemporal data analytics for renewable energy applications

Renewable energy domain	Techniques applied	Problem solved or discussed	References
Wind and solar	Generative adversarial networks (GANs)	A data-driven method is proposed for renewable energy by embedding interpretable manifold space in controllable GAN	[13]
Ensemble of renewable energy systems	e-MFTS (evolving Multivariate Fuzzy Time Series) and TEDA (Typicality and Eccentricity Data Analytics) framework	Forecasting in renewable energy systems due to uncertainties in data distribution	[14]
Solar forecasting	Various ML and statistical based methods	Reviewing solar forecasting dependence on atmospheric sciences	[15]
Solar, wind, hydro	Artificial neural networks (ANNs), fuzzy logic, hidden Markov models, classic ML techniques	Reviewing AI-based prognostic and maintenance of renewable energy systems	[16]
Wind and solar	RNNs, CNNs, LSTMs, GRUs, etc.	This work reviews solar and wind energy forecasting methods based on DL techniques	[17]
IoT sensors used in all renewable energy domains	K-means clustering for spatial analysis and tiny aggregation—temporal data analysis	IoT-based spatiotemporal data fusion is proposed for low-level data in and data out fusion for real-time spatial IoT source aggregation	[18]
Solar energy	GPS geo-location pointing, DBSCAN algorithm, eMBT EV model simulation using SUMO	Examines the energy needs of electric minibus taxis in an urban paratransit system and investigates charging possibilities at formal and informal stops	[19]
On efficient usage for all renewable energy systems	JERICHO-E-usage dataset, big data analytics, and linear programming (LP)	Using the JERICHO-E dataset and time-series analysis got useful energy usage prediction patterns for energy system modeling	[20]
All types of renewable energy systems	Statistical methods like the likelihood method and regression for spatial panel Dobbin model (SPDM)	For green finance use for the development of renewable energy in China	[21]
Green innovation for all types of renewable energy systems	Superefficient network SBM model, exploratory spatial data analysis (ESDA), and spatial econometric model	Provides the answers to research questions on the innovations to promote the utilization of efficient and sustainable green and renewable energy in all of China	[22]

(continued)

Table 10.1 (continued)

Renewable energy domain	Techniques applied	Problem solved or discussed	References
Solar energy	Fuzzy analysis and mutual information to construct spectral graph convolution design	Forwards a graph modeling system for photovoltaic (PV) power predictions done for short durations	[23]
Solar and wind energy	Novel method based on Tucker tensor decomposition that also involves PCA	Provide an efficient and effective prediction and forecasting model for renewable energy systems	[24]

cells based on the context of the input data. The key components of an LSTM cell are the input gate, the forget gate, the output gate, and the cell state. The input gate controls how much new information is added to the cell state, the forget gate controls how much old information is retained or forgotten from the cell state, and the output gate controls how much of the updated cell state is used to generate the output. The cell state is the memory of the LSTM cell, which is updated based on the input data and the state of the gates. LSTM networks have been widely used for various tasks, such as language modeling, speech recognition, machine translation, and video analysis. They have been shown to outperform traditional RNNs in tasks that require modeling long-term dependencies in sequential data.

LSTM networks are particularly useful for analyzing spatiotemporal renewable energy data because they can effectively model long-term dependencies in sequential data. This is important because renewable energy production is often influenced by weather patterns and other factors that change over time. In the context of renewable energy applications, spatiotemporal data analytics using LSTM networks can be used to analyze and predict renewable energy production and consumption patterns. For example, consider a wind farm consisting of multiple turbines located in different geographical locations. By collecting spatiotemporal data such as wind speed, direction, and temperature, we can train an LSTM network to predict the power output of each turbine at different time intervals. This information can be used to optimize energy production by adjusting the orientation and speed of the turbines, as well as to forecast future energy demand. Similarly, for solar energy applications, an LSTM network can be trained on spatiotemporal data such as solar irradiance, temperature, and cloud cover to predict the output of a solar panel array. This information can be used to optimize the orientation and tilt of solar panels, as well as to forecast future energy production. Figure 10.2 gives a pictorial depiction of LSTM. Figure 10.3 shows how an LSTM network can be used with spatiotemporal data for renewable energy applications. In Fig. 10.5, as you can see, firstly, the spatiotemporal data is transformed, decomposed, and normalized using various data optimization techniques. Then that optimized data is applied to a DL–input layer, and after that, it goes through 4-LSTM layers and then through a dense DL layer; the

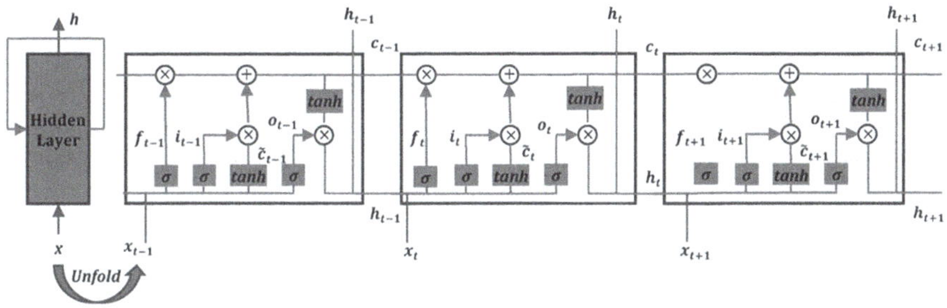

Fig. 10.2 Generic LSTM architecture

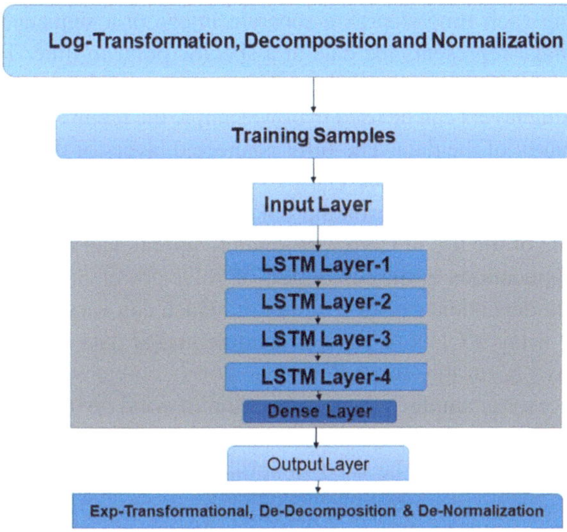

Fig. 10.3 Generalized pictorial description of LSTM usage on spatiotemporal renewable energy data

network of LSTM layers and dense layer can use mean square error (MSE) for training or any other method. After that, the output prediction about the usage and application of renewable energy can be ascertained.

2. *Convolutional Neural Networks (CNNs):* These are a type of neural network commonly used for image- and video-processing tasks. The architecture of a CNN consists of several layers, including convolutional layers, pooling layers, and fully connected layers. In the convolutional layers, the input image is convolved with a set of learned filters, which extract features from the image. The pooling layers downsample the feature maps to reduce their dimensionality, while the fully connected layers use the extracted features to classify the input image into one of several categories. The training of a CNN involves using a large set of labeled images to adjust the weights of the filters and fully connected

layers so that the network can accurately classify new images. CNNs have been shown to achieve good performance on a variety of computer vision tasks, including object detection, image segmentation, and facial recognition. They are also used in natural language processing tasks, such as text classification and sentiment analysis.

CNNs are well suited for spatiotemporal data analytics in renewable energy applications, as they can effectively capture patterns and trends in both the spatial and temporal dimensions of the data. In renewable energy applications, spatiotemporal data may include information about solar radiation, wind speed, temperature, and other meteorological variables at different locations and time intervals. These data can be used to predict renewable energy output or to optimize renewable energy systems. CNNs can be used to process spatiotemporal data by treating each time step as a separate image in a sequence of images, where each image represents the data at a specific point in time. The convolutional layers of the CNN can learn filters that capture spatial patterns in the data, while the pooling layers can be used to downsample the feature maps and reduce the dimensionality of the data. The fully connected layers of the CNN can then be used to predict the renewable energy output or to optimize the renewable energy system based on the spatiotemporal data.

One example of the use of CNNs for spatiotemporal data analytics in renewable energy applications is the prediction of solar power output. Solar power output is highly dependent on solar radiation, which can vary greatly over time and space. By using a CNN to process spatiotemporal data on solar radiation, researchers have been able to predict solar power output with high accuracy accurately. Another example is the optimization of wind turbines based on spatiotemporal data on wind speed and direction, where CNNs can be used to learn patterns in the data that can be used to optimize the performance of the wind turbines [17].

3. *Big Data Analytics (BDA):* Big data analytics plays a crucial role in spatiotemporal data analytics for renewable energy applications. Spatiotemporal data refers to data that is both spatially and temporally referenced, such as data collected by sensors placed on renewable energy sources like wind turbines, solar panels, or hydroelectric plants. The sheer volume of data generated by these sensors can be massive, making traditional data analysis methods inadequate. Big data analytics can help in processing and analyzing large volumes of spatiotemporal data, enabling faster and more accurate insights. This can lead to improved performance and efficiency of renewable energy sources.

   The first step in spatiotemporal data analysis is to collect the data from the sensors and store it in a database. This data can be preprocessed to remove any inconsistencies or errors. The next step is to apply machine learning and statistical analysis techniques to extract useful insights from the data. These techniques can include clustering, regression analysis, and time series analysis. In the case of renewable energy applications, big data analytics can be used to optimize the performance of renewable energy sources. For example, wind turbines can be optimized to produce maximum energy output based on wind speed and direction data collected by sensors placed on the turbines. Similarly, solar panels can be optimized to maximize energy output based on solar radiation data collected by sensors placed on the panels. Furthermore, big data analytics can be used to predict energy production and demand. This can help grid operators plan and manage the supply of renewable energy to the grid. For example, predictive analytics can help grid operators anticipate when energy demand will be high and adjust the supply of renewable energy accordingly [1, 18]. Figure 10.4 gives a graphical description of how big data and ML techniques can be applied for renewable energy applications of spatiotemporal data analytics.

4. *Generative Adversarial Network (GAN):* GANs are a type of deep learning model that is used in various applications, including spatiotemporal data analytics for renewable energy. GANs' neural network architecture consists of two main components: a generator and a discriminator. The generator generates fake data samples that resemble real data, while the discriminator tries to distinguish between fake and real data. The two components are trained together in a process called adversarial training, where the generator learns to generate better fake data by fooling the discriminator, while the discriminator learns to become better at distinguishing between fake and real data. During training, the generator takes random noise as input and generates a data sample, which is then fed into the discriminator along with real data samples. The discriminator then tries to identify which samples are fake and which are real. The feedback from the discriminator is used to update the generator, so it can generate more convincing fake data that can fool the discriminator. Over time, the generator learns to generate more realistic data samples while the discriminator becomes better at identifying fake data. In the end, the generator produces data that is similar to the real data, and it becomes difficult for the discriminator to tell the difference.

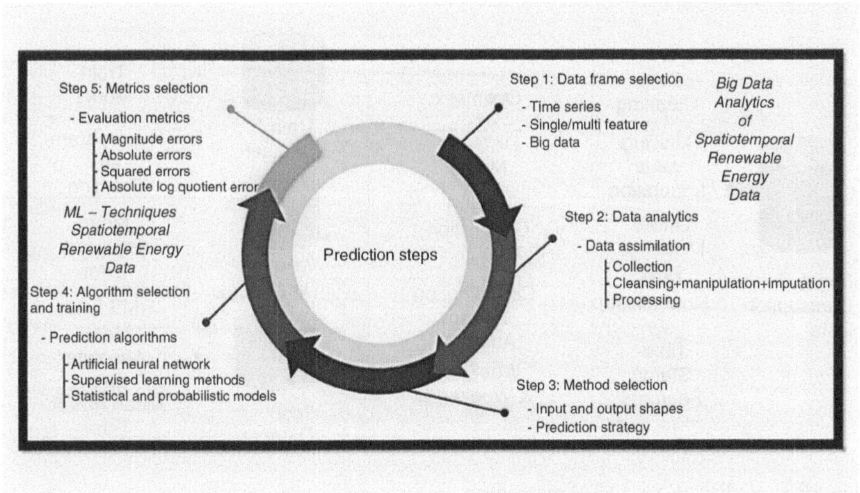

Fig. 10.4 Big data and ML techniques for spatiotemporal renewable energy data analysis

In this context, GANs can be used to generate realistic spatiotemporal data for renewable energy sources, such as solar and wind power. One of the main challenges in renewable energy applications is the variability of energy production over time and space. GANs can be used to generate realistic spatiotemporal data that can be used to model this variability and optimize energy production. GANs can be trained on the preprocessed data to generate realistic spatiotemporal data for renewable energy sources. GAN is made up of two neural networks. A generator network generates new data samples, and the discriminator network evaluates the realism of these samples. The generated spatiotemporal data can be used for various data analysis tasks, such as forecasting energy production, identifying patterns and trends, and optimizing energy production. The generated data can be compared to real-world data to validate the performance of the GAN and ensure that the generated data is realistic [3].

An example of the application of CNN-LSTM can be seen in [25], and CNN-LSTM is used to design and implement a deep learning-based framework for building energy consumption forecasts. The kCNN-LSTM is a deep learning framework which operates on the energy consumption data that is recorded at some predefined intervals to provide precise building energy consumption predictions. Figure 10.5 gives the kCNN-LSTM building energy consumption prediction model as formulated and designed [25].

Now, we present a list of metrics that are generally used with the above-discussed methodologies and technologies for analyzing spatiotemporal data for renewable energy:

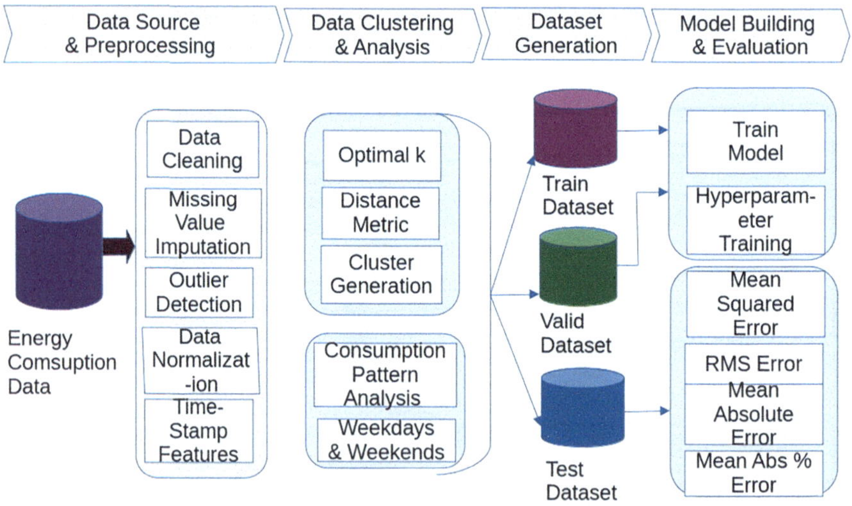

Fig. 10.5 Applied kCNN-LSTM mechanism

- *Capacity Factor (*CF*):* This metric represents the actual output of a renewable energy project divided by the maximum possible output over a given time period. It is usually expressed as a percentage and is a measure of how efficiently a project is generating energy.
- *Levelized Cost of Energy (LCOE):* This metric represents the total cost of building, operating, and maintaining a renewable energy project, divided by the total amount of energy it generates over its lifetime. It is usually expressed as a dollar value per unit of energy (e.g., cents per kilowatt-hour) and is a measure of the cost-effectiveness of a project.
- *Power Density:* This metric represents the amount of power that can be generated per unit of land area or water surface area. It is usually expressed as watts per square meter (W/m^2) or megawatts per square kilometer (MW/km^2) and is a measure of the spatial efficiency of a project.
- *Capacity Utilization Factor (CUF):* This metric represents the actual output of a renewable energy project divided by its maximum possible output over a given time period, taking into account the intermittent nature of renewable energy sources. It is usually expressed as a percentage and is a measure of how well a project is able to use the available resources.
- *Renewable Energy Fraction (REF):* This measure quantifies the proportion of energy consumed that originates from renewable sources, typically presented as a percentage. It serves as an indicator of the extent to which renewable energy contributes to the overall energy composition.
- *Carbon Intensity:* This metric represents the amount of carbon dioxide emissions associated with generating a unit of energy from a renewable energy source. It is usually expressed as grams of CO_2 equivalent per kilowatt-hour (gCO_2e/kWh) and is a measure of the environmental impact of renewable energy.

These are just a few examples of the metrics that can be used to analyze and compare spatiotemporal data for renewable energy. The selection of metrics relies on the distinct objectives of the analysis and the attributes of the renewable energy sources under examination.

10.4 Research Directions for Future Applications

Based on the literature survey done for this chapter, the following future research directions can be worked on by budding researchers in the domain of renewable energy research by using spatiotemporal data analysis:

- Developing more precise and reliable spatiotemporal prediction models for renewable energy generation, load demand, and weather conditions using machine learning and deep learning algorithms. These models can help optimize the integration of renewable energy sources into the grid and improve the grid's stability and efficiency.

- Investigating the use of federated learning and other privacy-preserving techniques for spatiotemporal data analytics to protect sensitive data while still allowing for the sharing and collaboration of data between different organizations.
- Exploring the potential of digital twins, AI, and cyber-physical systems for building energy management and optimization in the context of renewable energy systems. These technologies can help improve energy efficiency, reduce energy waste, and optimize energy use in buildings and other structures.
- Developing new spatiotemporal data fusion techniques for integrating data from different sources and sensors, including IoT devices and big data analytics platforms. The use of these techniques can help improve the precision, accuracy, and reliability of renewable energy forecasting, which in turn can lead to more informed decision-making.
- Investigating the potential of renewable energy systems for transportation, including electric vehicles and solar-powered charging stations. This area of research can help promote the adoption of renewable energy sources in transportation and reduce carbon emissions.
- Studying the economic and financial aspects of renewable energy projects, including financing options and investment strategies. The findings of this research can be used to inform the development of policies and regulations that will promote the use of renewable energy, which can help to reduce our reliance on fossil fuels and improve the sustainability of the energy sector.

10.5 Summary

This book chapter is based on the use of spatiotemporal data analytics for application areas concerning renewable energy such as wind energy, solar energy, hydropower, and also sustainable usage of energy in buildings, houses, industries, and other large critical infrastructure. First and foremost, discuss the basics of spatiotemporal data and its analysis using various available software and mathematical tools. In the introduction section, we dwelled upon the use of spatiotemporal data analytics for renewable energy applications. A very comprehensive and detailed literature review on the recent state-of-the-art research done in the domain of the chapter topic is provided. The methods and technologies used in the applications and implementations for the prediction and forecasting of efficient and proper usage of renewable energy for a sustainable and brighter energy future for humanity are explored. In the final section, we provide the readers with a sneak peek into the future research directions that this research domain could take. In summary, this chapter offers an extensive and meticulous examination of the analysis and utilization of spatiotemporal data in renewable energy, focusing on its efficiency, sustainability, and effectiveness.

References

1. Alam, M. M., Torgo, L., & Bifet, A. (2022). A survey on spatio-temporal data analytics systems. ACM Computing Surveys, 54(10s), 1-38.
2. Song, J., Zhang, L., Xue, G., Ma, Y., Gao, S., & Jiang, Q. (2021). Predicting hourly heating load in a district heating system based on a hybrid CNN-LSTM model. Energy and Buildings, 243, 110998.
3. Li, Y., Li, J., & Wang, Y. (2021). Privacy-preserving spatiotemporal scenario generation of renewable energies: A federated deep generative learning approach. IEEE Transactions on Industrial Informatics, 18(4), 2310-2320.
4. Gu, C., & Li, H. (2022). Review on deep learning research and applications in wind and wave energy. Energies, 15(4), 1510.
5. Ahmad, T., Zhang, H., & Yan, B. (2020). A review on renewable energy and electricity requirement forecasting models for smart grid and buildings. Sustainable Cities and Society, 55, 102052.
6. Ifaei, P., Nazari-Heris, M., Charmchi, A. S. T., Asadi, S., & Yoo, C. (2022). Sustainable energies and machine learning: An organized review of recent applications and challenges. Energy, 126432.
7. Neumann, F., & Brown, T. (2021). The near-optimal feasible space of a renewable power system model. Electric Power Systems Research, 190, 106690.
8. Lv, J., Zheng, X., Pawlak, M., Mo, W., & Miśkowicz, M. (2021). Very short-term probabilistic wind power prediction using sparse machine learning and nonparametric density estimation algorithms. Renewable Energy, 177, 181-192.
9. Kumari, P., & Toshniwal, D. (2021). Long short-term memory–convolutional neural network based deep hybrid approach for solar irradiance forecasting. Applied Energy, 295, 117061.
10. Kezunovic, M., Pinson, P., Obradovic, Z., Grijalva, S., Hong, T., & Bessa, R. (2020). Big data analytics for future electricity grids. Electric Power Systems Research, 189, 106788.
11. Agostinelli, S., Cumo, F., Guidi, G., & Tomazzoli, C. (2021). Cyber-physical systems improving building energy management: Digital twin and artificial intelligence. Energies, 14(8), 2338.
12. Raghutla, C., Shahbaz, M., Chittedi, K. R., & Jiao, Z. (2021). Financing clean energy projects: New empirical evidence from major investment countries. Renewable Energy, 169, 231-241.
13. Dong, W., Chen, X., & Yang, Q. (2022). Data-driven scenario generation of renewable energy production based on controllable generative adversarial networks with interpretability. Applied Energy, 308, 118387.
14. Severiano, C. A., e Silva, P. C. D. L., Cohen, M. W., & Guimarães, F. G. (2021). Evolving fuzzy time series for spatiotemporal forecasting in renewable energy systems. Renewable Energy, 171, 764-783.
15. Yang, D., Wang, W., Gueymard, C. A., Hong, T., Kleissl, J., Huang, J., ... & Peters, I. M. (2022). A review of solar forecasting, its dependence on atmospheric sciences and implications for grid integration: Towards carbon neutrality. Renewable and Sustainable Energy Reviews, 161, 112348.
16. Saleem Afridi, Y., Ahmad, K., & Hassan, L. (2021). Artificial Intelligence Based Prognostic Maintenance of Renewable Energy Systems: A Review of Techniques, Challenges, and Future Research Directions. arXiv e-prints, arXiv-2104.
17. Alkhayat, G., & Mehmood, R. (2021). A review and taxonomy of wind and solar energy forecasting methods based on deep learning. Energy and AI, 4, 100060.
18. Fawzy, D., Moussa, S., & Badr, N. (2021). The spatiotemporal data fusion (STDF) approach: IoT-based data fusion using big data analytics. Sensors, 21(21), 7035.
19. Abraham, C. J., Rix, A. J., Ndibatya, I., & Booysen, M. J. (2021). Ray of hope for sub-Saharan Africa's paratransit: Solar charging of urban electric minibus taxis in South Africa. Energy for Sustainable Development, 64, 118-127.
20. Priesmann, J., Nolting, L., Kockel, C., & Praktiknjo, A. (2021). Time series of useful energy consumption patterns for energy system modelling. Scientific Data, 8(1), 148.

21. Tang, X., & Zhou, X. (2023). Impact of green finance on renewable energy development: A spatiotemporal consistency perspective. Renewable Energy.
22. Liu, P., Zhang, L., Tarbert, H., & Yan, Z. (2022). Analysis on spatio-temporal characteristics and influencing factors of industrial green innovation efficiency—from the perspective of innovation value chain. Sustainability, 14(1), 342.
23. Cheng, L., Zang, H., Ding, T., Wei, Z., & Sun, G. (2021). Multi-meteorological-factor-based graph modeling for photovoltaic power forecasting. IEEE Transactions on Sustainable Energy, 12(3), 1593-1603.
24. Corizzo, R., Ceci, M., Fanaee-T, H., & Gama, J. (2021). Multi-aspect renewable energy forecasting. Information Sciences, 546, 701-722.
25. Somu, N., MR, G. R., & Ramamritham, K. (2021). A deep learning framework for building energy consumption forecast. Renewable and Sustainable Energy Reviews, 137, 110591.

Chapter 11
Environmental Spatiotemporal Data Analytics

Shubhangi Tidake, Bandana Mahapatra, and Suchit Subodh Mishra

Abstract Environmental spatiotemporal data analytics (ESTDA) is a field that combines environmental science, data science, and geographic information systems to explore the relationship between environmental phenomena and their spatial and temporal variability. ESTDA has been used to address a wide range of environmental issues, such as climate change, pollution, biodiversity loss, and natural disasters. The goal of this field is to identify patterns, trends, and anomalies in environmental data that can help scientists and policymakers make informed decisions. ESTDA relies on a variety of analytical techniques, including statistical models, machine learning algorithms, remote sensing, and spatial analysis. These techniques allow researchers to extract meaningful information from large and complex datasets, including environmental monitoring networks, satellite imagery, and citizen science data. By analysing these data, ESTDA can provide insights into the drivers of environmental change, the impacts of human activities on the environment, and the effectiveness of environmental policies and management strategies. Overall, ESTDA has the potential to improve our understanding of environmental systems and inform more effective environmental decision-making. However, it also faces a number of challenges, such as data quality and availability, computational limitations, and need for interdisciplinary collaboration. Addressing these challenges will be crucial for the continued advancement of ESTDA and its potential to contribute to sustainable development and conservation efforts. The chapter aims at addressing the overall concept of ESTDA, its issues, and challenges.

Keywords Spatiotemporal data · Environmental analysis · Trajectory data segmentation · Decision-making

---

S. Tidake · B. Mahapatra (✉)
Department of Data Science, Symbiosis Skills and Professional University, Pune, Maharashtra, India
e-mail: bandana.mahapatra@sspu.ac.in

S. S. Mishra
Jazzee Technologies Limited, London, UK

J. A et al. (eds.), *Spatiotemporal Data Analytics and Modeling*, Big Data Management, https://doi.org/10.1007/978-981-99-9651-3_11

11.1 Introduction

The concept of spatial signifies space, whereas temporal refers to time. "Spatiotemporal" is a technique related to analysing the data belonging to both space and time domain. It majorly describes the methodology of data analytic conducted in each specific location and time, for example, water vehicle transportation in a given geographical area over time. The application areas of spatial-temporal reasoning are majorly focused on solving multifaceted problems by analysing or predicting the manner in which objects change themselves with respect to both space and time.

The domain of spatiotemporal based analytics of a given feature-based data is currently undergoing a technological revolution with much focus over research due to the evolvement of powerful computational machines, various high computing processors such as graphic processing units (GPUs) that can process huge quantity of voluminous data using analytics techniques for "big data analytics". The spatiotemporal databases offers a variety of data collated from space as well as time domain which may express an event being occurred at a particular location at a specific given time.

The various domain areas where spatiotemporal based data analytics can be implemented are biological science, ecological science, meteorology, medical science, logistics, and forestry. Spatial-temporal based feature and fact representation is challenging as an object being modelled in space has n number of directions, whereas time can only move forward. The combination of both the aspects for representing and understanding dual factors, i.e., space and time, makes the analytical part all the more complex.

The concept of generating synthetic data can vary based on the way continuum is expressed, i.e., a zip code. The time aspect may also be creating confusion based on weather calculated measurements based on secs, mins, hrs, days, or years. The main goal of spatiotemporal analytics is forecast and information extraction from diverse as well as complicated data collected.

11.1.1 Temporal Databases vs. Spatial Databases

A temporal database is a database that holds the stock of data pertaining to time-based occurrences. The temporal databases can be in any of the following forms: uni-temporal, bi-temporal, or tri-temporal. The temporal aspects consist of accurate time, transaction time, or decision time where valid time denotes the time period considered when the events occur, transaction time is the time recorded within the database, while the decision time may be defined as the time during which the appropriate conclusions were made regarding a given fact.

A temporal database consists of associated data collected across all the three tenses of time, i.e., past, present, or future. It helps to form connection between data accuracy with respect to the time of its entry.

On the other hand, a spatial database supports repositories of data and its related queries based on data collected over a geometric space.

The repository as well as retrieval of spatial and temporal data is essentially needed for application areas such as geographical information systems (GISs), environmental information systems (EISs), and neuro-informatics, whereas the spatial-temporal database deals with data representing mobile objects [1].

11.1.2 What Is Spatial Temporal Reasoning

Spatial-temporal reasoning deals with the various methods that are used by a robot equipped with AI for collation of objects within a space. This understanding and reasoning are achieved by picturising a spatial pattern which consists of visualising stepwise process on how objects make various patterns.

Spatial-temporal contributes to the concept of cognitive psychology, which looks at the aspects regarding how a mind is able to process its knowledge base equipped with data based on time and space.

Spatial-Temporal Reasoning Examples: These test cases are practically implemented in technical engineering and military recruitments, computational biology, as well as document interpretation [2].

11.2 Environmental Data

The technical advancement in the fields of Internet of Things (IoT) and geolocation technology has resulted in an increase in the number of connected objects that are location aware, including vehicles, vessels, as well as biologically mobile species such as humans and animals.

Currently, smartphones are embodied with GPS that is not just a mere emerging trend, but almost an essential feature. Such mobile pervasive sensor-equipped technologies support consistent production of heavy volumes of geo-based data series, which have no restriction over time and space, as well as promote the connectivity existing within physical objects and their surroundings. The conventional wireless sensor-based network essentially needs a huge number of deployed sensors that can ensure a wide area coverage for a large-scale and fine-grained sensing, which is productively impractical and monotonous.

The region under consideration had 30 stations localised in Paris, whereas the rest were scattered around covering the whole area. This problem can be addressed by the novel model named as mobile crowdsensing (MCS) [1, 2] that facilitates the data-acquiring process via a multi-sensory box, along with a mobile device that can be watchful of large-scale phenomena. In the MCS environment, the participants have multiple sensors along with a GPS-embedded mobile device. The devices can rotate freely within the boundary region of monitoring to get hold of samples as well as collate and convey the collected data to the monitoring centre for covering the observation conducted. This leads to an uninterrupted production of huge data based on geo data series. The special trait of this model is a unique combination of geolocation-based observations and measurements of the observed phenomena.

Ubiquitous computing majorly concerns about understanding of user mobility from GPS and sensory-based data. Intensive and extensive researches have been conducted till date in the field of sensory data mining in order to extract important information from raw data. The chapter majorly focuses on a considerable number of related studies and surveys conducted to analyse and manage enriched trajectory data.

Primarily, a trajectory is defined as a multivariate time series in which one dimension represents the spatial positions. In contrast, enriched trajectories are trajectories enriched with other facts such as the context or the whereabouts of the mobile object. The augmented trajectories when developed in context to MCS are constructed GPS tracks with temporal measurements complemented with other contextual information about the moving object. Hence, the state-of-the-art capturing of this kind of enriched trajectory data as shown in Fig. 11.1 comprises five components:

1. Trajectory data pre-processing
2. Trajectory data segmentation
3. Activity recognition
4. Generic machine learning algorithms
5. Moving object databases and warehousing

The structured design of enriched trajectory-based data mining shown in Fig. 11.1 was inspired from the survey of Zheng [3] and gives an overall comparison of the state-of-the-art techniques, which provides information from raw trajectory data. In his survey, Zheng has identified various connections and differences existing among the techniques mentioned in the literature. The survey has put forth a paradigm for trajectory-based data mining. They proposed a model that allows the raw trajectory data to convert into productive information by undergoing various steps, which can be categorised into four groups:

1. Derivation of trajectory data
2. Trajectory data pre-processing

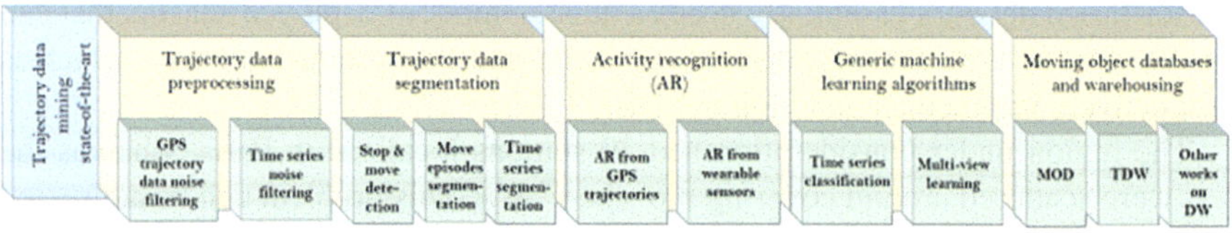

Fig. 11.1 Structure of state-of-the-art chapter

3. Trajectory data management
4. Various mining tasks: trajectory pattern mining and trajectory classification

11.3 Trajectory-Based Data Pre-processing

The mining approach may be defined as the overall approach which combines the temporal data along with the missing GPS data. This chapter focuses on conducting a comprehensive review of related studies for enrichment of trajectory mining in terms of pre-processing, segmentation, activity recognition, management, and warehousing as shown in Fig. 11.1. The chapter also highlights the differences between trajectory data mining and time series mining as the combination of these two components contributes to enriched trajectories.

The trajectory-based data pre-processing proposed by Zheng [3, 4] recounts basic techniques required to process the trajectory-based data before starting the mining tasks. These techniques are based on noise filtering, stay point detection, and trajectory segmentation. The section mainly highlights the noise filtering, while trajectory segmentation and the concept of stay point detection are discussed in Sect. 11.3. The concept of noise filtering is mainly to remove the noise factor from the data that may cause bias in the mining. The sensory data collected suffers from inaccuracy and noise due to technical errors such as poor receiving of positional signals in urban areas. Filtering out these imperfections aims at improvising the quality of the data for performing mining of the data.

Elnahrawy and Nath in their article [5] have given a comparative analysis between the various sources of errors. The article has broadly classified the errors into systematic errors (bias) and random errors (noise). The systematic errors occur due to changes in the operating conditions (e.g. temperature, humidity) or expiration of the sensors participating. These glitches can be corrected due to calibration as done in the article [5].

The random errors include the unpredictable variation arising due to variations from one measurement to another occurring due to generation of noise from various external sources and environmental factors. The chapter aims at identifying random errors and various measures to reduce their effects on sensor readings that may influence the reading precision. The primary intent is to overcome the issues related to data quality of the collected data from various sensors via usage of rule-based models or mathematical techniques. These techniques focus on both trajectory data and time series data.

11.3.1 GPS Trajectory Data Noise Filtering

Getting erroneous data due to inaccurate readings taken from GPS devices is quite common due to several technical reasons such as poor receiving signals and technical device glitch. The issues are yet to be solved and hence have attracted a lot of attention from the researchers who have conducted several studies for filtering such noise. The universally accepted approach focuses on heuristic methods that potentially remove noise points directly from the trajectory, as conducted in GeoLife [6] and Scikitmobility [7]. These heuristic approaches first calculate the speed of the trip between consecutive points in the trajectory, which is derived over the distance between the time spent between a point and its successor. When the speed exceeds certain threshold as defined by the user (e.g. 300 km/h), the segment of the two consecutive points is cut off. However, an article published by Hendawi et al. [8] suggests that every data that qualify as noise in GPS data need not be discarded, rather should be analysed deeply as they may reveal some vital information about the place where the user is present in indoor spaces. Based on this given hypothesis, the researcher Hendawi and his team focused on discovering various patterns of noisy GPS data and connecting them to specific locations. For example, it may be possible to "conclude that a driver is coming across a tall building or a forest based on the pattern of noise received in the GPS readings". Hence, by tracking the noise pattern, the researchers may assume the height of surrounding buildings, the existence of urban constructions, and the existence of tunnels in the geospatial data.

11.3.2 Time Series Noise Filtering

The detection of noise and outliers in time series data typically requires dedicated techniques of different types. Gupta et al. [9] and Blázquez-García et al. [9] have conducted an extensive and structured survey over the various latest techniques that have been widely used in the time series outlier detection domain. The authors in the article discuss numerous existing approaches that range from statistical methods (such as autoregressive (AR) models and ARIMA models) to more complex and advanced methods (such as deep learning-based approaches). The authors in the article [10] have instigated other researchers to conduct more studies over literature that aim at detecting and filtering the noise from time series data. In similar lines, Palshikar [11] proposed a technique based on peak detection (aka, spike detection) for cleaning outliers from given time series data. The peaks in the data may be defined as an occurrence of a spike in the values of time series. Such spikes are easily identifiable visually in a short time series, but it is not quite obvious to be detected automatically in any time series. Hence, the proposed models have introduced a formal notion of a peak in time series, and further various algorithms are proposed for detection of peak. They further defined the local peak as the point which is the local maximum within a window size, i.e. the value of this point is different to the other values existing in that window. The identification of peaks needs a user-defined threshold of the window average to differentiate between local outliers and real values. Currently, the need of time is automatising of the selection of threshold based on the window mean value.

11.4 Trajectory Data Segmentation

In real-world applications, the trajectory data needs to be segmented into coherent segments for further data analysis. The segmentation facilitates the mining of rich knowledge from the trajectory data, e.g. inference of stay locations. It may be categorised as the first step of semantic enrichment of data. The various approaches and methods proposed for trajectory data segmentation have made the mobility structure quite rich. The proposals are developed over specific application domain aims at achieving the research objective, e.g. deriving the stay area of a moving object, or detection of transportation mode transitions within a trajectory, aiming towards achieving objectives of the specific application domain. This modelling approach is quite similar to the time-geography representation proposed by Hägerstrand [12], which may be described as the first successful proposal for providing a conceptual approach for representing and analysing human activity in space and time.

Figure 11.1 is also motivated from the three-dimensional representation of a trajectory modelled as per the framework of time-geography. The concept of time-geography had evolved before the emergence of smartphones, which focus on modelling the constraints of human activity in the trajectory of space and time. A massive collection of data based on location reinforces conduction of analytical approach over this paradigm.

However, this approach has raised a new challenge with respect to data pre-processing, among which trajectory reconstruction, stop and move detection, and activity labelling are the primary factors under consideration. Also, via incorporation of the concept of time-geography into real-life measurements, defining the scope is very essential with minute level of detail [13].

This section of the chapter primarily focuses on three types of segmentation methods for trajectory data mentioned as:

1. Location-derived tracking of stop and movement
2. Movement-based episode segmentation: speed and time series-based trajectory segmentation
3. General model for change-point segmentation in time series data

11.4.1 Stop and Move Detection

This trajectory segmentation is mainly application driven, while trajectory focuses on detecting movements having several trips, such as relay from starting point A, passing through point B, then point C, and finally to an end point D as described in Fig. 11.2. Performing segmentation of such a trajectory is splitting it into segments where the moving objects are stationary called the stop segments from the moving object which is actually moving called the moving segments.

The approach is called as stop and move segment detection, where a move segment is always between two stop segments.

The research-based problem statement that can be considered by researchers is identification of stop and move segments. The article contributed by Parent et al. [14] discusses the latest approaches proposed for locating stops within a trajectory. The authors in the article aim at identifying the existing criteria and assumption for detecting stop-based segmentation. Quite a few of these assumptions are derived from the spatial distance and temporal duration as mentioned in the article [15] contributed by Zheng et al., where the stop events are calculated with respect to various sequences of consecutive GPS points. Here, the spatial distance is lower than the threshold, whereas the temporal duration is quite higher than the other threshold. The primary contribution to stay point detection was proposed in an article contributed by Li et al. [15], which developed an algorithm for identifying stop-based segments. In the work, further algorithm was proposed focused on finding the stay point (from P5 to P8 with P5 as an anchor point), as well as examining the successor points from P6. This can be explained further as, if the distance between P6 and P9 is within the threshold, P9 will be added to the stay point.

The various methods developed to identify the stop segments are created by using a combination of spatiotemporal indicators. An article contributed by Yan et al. [16] has proposed a method to identify the stop and move segments by considering various spatiotemporal criteria such as position density factor, velocity, and direction. Various other lines of methods designed for discovering the stop segments are incepted via using only coordinates and time stamps captured from continuous GPS trajectory, excluding the feature of velocity. In their article, Gong et al. [17] proposed a two-step approach for identification of stops based on the variant of

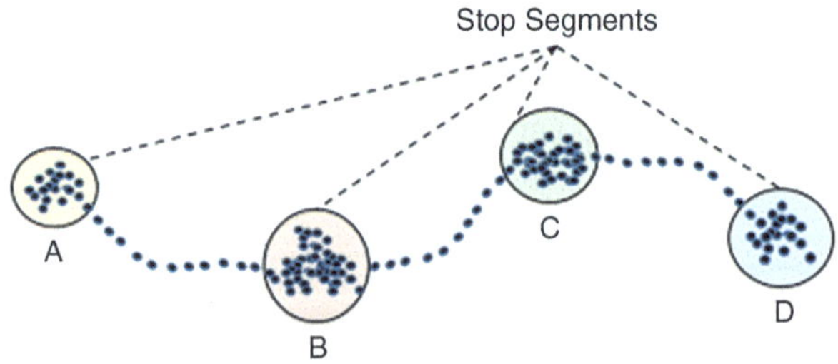

Fig. 11.2 Stop and move segments in a trajectory

DBSCAN [18] algorithm and support vector machine (SVM). The proposed variant, named DBSCAN-TE (Density-Based Spatial Clustering of Applications with Noise extended with Temporal and Entropy constraints), identifies the stop clusters as well as feeds the output to SVM, which could differentiate between activity stops (e.g. work, shopping) and non-activity stops (such as waiting for a green light, being stuck in a congestion).

11.4.2 Move Episode Segmentation

The various trajectory-based data can be segmented as per given specifications of stops and moves, for example, criteria of transportation means for segmenting trajectory data. Few of the research works that contributed include Zheng et al. [19] who used speed and acceleration as criteria for segmentation of trajectory data and trace changing points. The concept is built upon facts such as human objects move between two transportation modes, where authors have considered the starting and ending of the walk segment as the change point for calculation. Hence, the first distinguishing criteria are walk point and non-walk points conducted by considering a loose upper bound of velocity and acceleration. Next is construction of walk and non-walk segments via consecutive walk and non-walk points, where the beginning and end points of each walk segment may be considered as probable change points for segmenting the trip. Similar approach was developed by Liao et al. in their article where they used the Gaussian mixture model based on velocity, where the authors classified the trajectory data into three speed ranges: walking, low speed, and high speed. Hence, a new segment was created whenever there was a variation over the speed range. Going further from using fixed spatial or temporal threshold, Bonavita et al. [19] offered a general model which identifies users' mobility and checks the segmentation threshold that could fit into their mobility features. The proposed approach allows avoiding the possible input parameters as well as adapts the thresholds to user's trajectory data. Similar model proposed by Yan et al. [20] in their article aims at segmenting the data based on semantic information such as public transport and networks. In case where the person uses a bike path, it can be determined that the transportation mode is either walking or bike, but definitely neither bus, nor train, nor car. Apart from this, the location of bus stops and train stations indicates whether the travelling means is via bus or train.

11.4.3 Time Series Segmentation

Time series segmentation deals with the issue of segmenting data into coherent segments. For example, based on the application requirements, the segmentation can be typically for improvement of identifying the human activity via data segmentation that segregates the data into non-overlapping segments, where each segment represents a single activity. The existing change point detection approaches discussed may be categorised with respect to sensor-based technology, kinds of activities, segmentation methods, etc. The survey conducted in the article by Aminikhanghahi et al. [21] enumerates, categorises, and compares the various methods proposed for detection of change points in time series in both batch and online modes. The choice of the method depends upon the desired outcome of the algorithm. The issue in activity segmentation is about identifying the change point detection in IoT. Several advanced researches based on wearable sensors [20], camera [2], or smart home [7] have been discussed in the state of the art related to collecting information, segmenting data, as well as understanding and inferring human activities. The problems of activity recognition and activity segmentation are hugely interconnected topics since the starting point of human activity recognition is into detection of the various transition points that label them with activities.

The time series graph depicts the change of temperature and activities over the time frame of 9 h. The plot highlights the change in activity over nine times. The plot covers the segment between two consecutive change points being referred to as an activity. Henceforth, the objective of change point detection is towards discovering such activity borders with a change point detection.

In Fig. 11.3, a sample of time series data has several change points that are depicted by the green dashed lines. The survey paper contributed by Aminikhanghahi et al. [6] provides an extensive review of supervised (e.g. decision tree, support vector machine) and unsupervised machine learning algorithms (e.g. cumulative sum, Bayesian model) that have been proposed to identify change point.

The supervised methods for identification of change points begin with a training set that is extensively used to learn the process of delineating the identified feature from the given input data. The data here is labelled via activity classes that is aggregated along with data collection or is given by an expert.

The unsupervised learning algorithms, over the contrary, were used for identifying change patterns in an un-labelled data. Considering that the data is time series, these approaches may be used for uncovering the time series-based transitions

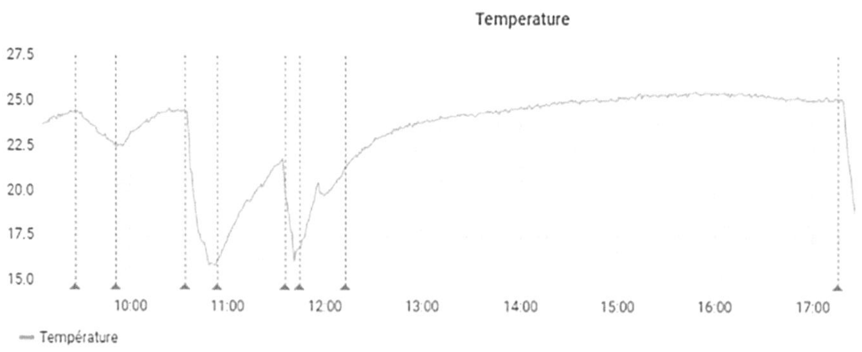

Fig. 11.3 Example of time series that contains several change points depicted by the green dashed lines

without any advanced intimation related to class labels or a training set. In an article contributed by Aminikhanghahi et al. [7], they have spotted two major unsupervised learning techniques used for detection of time series-based transition over unlabelled data mentioned as:

(a) Relative unconstrained least-squares importance fitting (RuLSIF)
(b) Bayesian change point detection (BCPD)

The outcome compared proves that the proposed Rul_SIF algorithm gives the better result for activity transition detection. But the challenge is the decision regarding choosing of appropriate values for RuLSIF parameters that may cause major notable changes with respect to point detection results. Similar contribution considered the real-time non-parametric change point detection, called SEP, that uses dissociation distance as a diverse methodology for detecting the change points within high-dimensional time series data [9]. The main aim of the proposal is extended on the research line towards density ratio change point detection algorithms, as well as introducing new unsupervised algorithms for identification of change point in time series data via dissociation distance metric.

Similar article contributed by Sadri et al. [21] proposes an offline change point detection method derived from information gain theory. According to the authors' idea, changes that alter the time series' mean can be found by counting change points. Using a moving window, the suggested approach can also spot changes in the time series' variance. The performance of change point identification, however, may be impacted by determining an advanced number of change points within the given time series.

Two alternative approaches to time series segmentation (unrelated to change point identification) are put out by Liu et al. [22]. Using a specific window size and sliding length, the data stream is explicitly split into a number of sub-sequences in the first way. However, the precision of the transition detection is directly impacted by the two parameters.

The second technique, called sensor event-based segmentation, separates the data stream into sub-sequences that each includes a specific number of sensor events. This method allows the window size to be dynamically adjusted to meet various recognition activities.

Gharghabi et al. [23] provide a domain-agnostic technique for multidimensional time series segmentation with regard to multivariate time series segmentation. The suggested method, fast low-cost unipotent semantic segmentation (FLUSS), takes advantage of the matrix profile structure first established in [24] to identify changes in the time series' shape.

To sum up the conversation, there are two ways to look at the studies of enriched trajectory data segmentation in the literature. The first one, which is based on GPS trajectory data, first splits the data into stop and move segments before looking at the speed's time series to identify transportation means' transitions inside move segments. It is true that speed has demonstrated success in dividing individuals' trajectories into several segments in accordance with the change in mode of transportation, as demonstrated in [23, 24], but speed is extremely susceptible to traffic and weather circumstances. The change in participant activities might not be recorded if driving speeds were on par with cycling during a traffic bottleneck. The second viewpoint is based on the segmentation of time series.

In these existing studies of change point detection methods for univariate time series, segmentation is taken up. While the proposal of Zheng et al. [7] demonstrated the highest performance using the RuLSIF method for activity transition identification, the results of change point detection are greatly influenced by the values specified for the RuLSIF parameters. Similar to this, setting the number of change points across the entire time series has an impact on the outcomes of change point detection in the proposal of Bonavita et al. [20], which can detect changes in the variance of the time series using a moving window.

Additionally, two distinct actions may show in the same window in the study of Zheng et al. [5], where the authors used a certain window size and sliding length to split the data. As a result, it might not be possible to determine the precise change point timestamp.

The approaches previously mentioned only apply to time series with a single variable. Although the work of Pappalardo et al. [8] can be applied to multidimensional time series, it makes the assumption that the reader is already familiar with the subset of time series dimensions that are important for identifying each change point.

The multivariate time series' dimensions do not contribute equally to the detection of activity shift in the setting of ambient crowdsensing, and the relative importance of each dimension is unknown beforehand.

11.5 Activity Recognition

Applications for recognising human activity are numerous and include smart home activities [8], daily human activities [22], and human mobility [20], in specific. This reflects a typical machine learning environment, where the benchmarks make extensive use of certain accessible datasets. We evaluate a few researches on activity detection using mobile sensing data in this part. We primarily concentrate on extrapolating activities from GPS trajectory data and wearable sensors.

11.5.1 Activity Recognition from GPS Trajectories

Using GPS-based trajectory data, various researches have begun to concentrate on activity identification in recent years. To understand the raw trajectory data (or its segments) and the mobility of the moving item, this type of challenge tries to tag the raw trajectory data (or its segments) with semantic labels, as shown in Fig. 11.4.

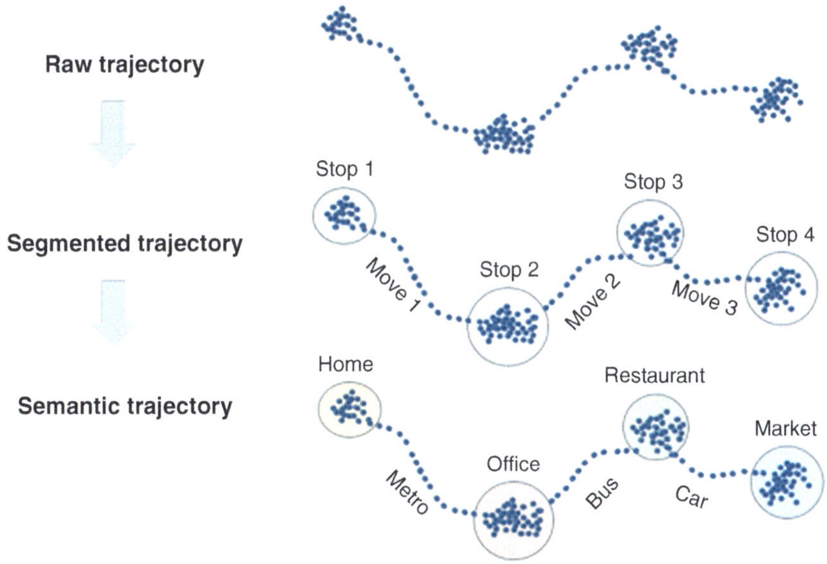

Fig. 11.4 Raw trajectory data to semantic trajectory

Numerous applications of the semantic information include delivering trip recommendations or finding the locations that a moving object frequents the most. Zheng [23] distinguishes between the applicability criteria for Fig. 11.4:

Generate semantic trajectory from raw trajectory data by adding a semantic label to raw trajectory data (or its segments), such as "transport modes inference" or "human activities detection".

His proposed diagram outlines three processes for activity identification from GPS trajectory:

1. Use a segmentation method to divide the data for the trajectory. As an example, divide the trajectory data into stopping- and movement-based segments, and then divide the move segments in accordance with changes in the mode of transportation.
2. For each segment (or point) of the trajectory, extract features.
3. Create a model that enables segment (or point) classification.

The general data flow for activity recognition from GPS trajectory data is shown in Fig. 11.5. Since a trajectory is primarily a sequence, information from local points (or segments) on the trajectory as well as the sequential patterns within close spaced points (or segments) can be combined using sequential based inference-driven models like dynamic Bayesian network (DBN), hidden Markov model (HMM), and conditional random field (CRF).

A more recent assessment by Mazimpaka and Timpf [88] examines several general techniques for mining trajectory data as well as their interrelationships. In line with Zheng's methodology, the authors claim that the majority of trajectory

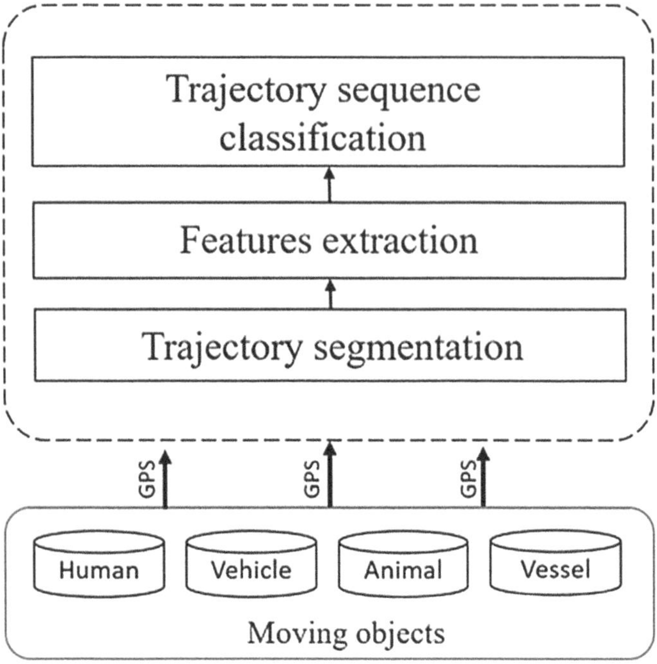

Fig. 11.5 General data flow for activity recognition from GPS trajectory data

classification algorithms use a conventional two-step methodology: first, they extract a collection of discriminative features, and then they use those features to train an already existing standard classification model.

A three-step trajectory data mining methodology based on machine learning methods is proposed and evaluated by Rehrl et al. in their publication [24]. The authors concentrate on issues with stopping locations in vehicle trajectory detection and classification.

The suggested method outlines three mining steps: stop detection, feature extraction, and classification of stoppage segments. The trajectory is first segmented into clusters of stay points (as discussed in Sect. 11.3.1 for stay point detection), and then, after extracting 14 attributes from each stop, the authors divide the identified stops into two groups: stops that are significant to traffic and stops that are not. The research community has produced several thorough studies based on machine learning approaches in the field of transportation mode detection. A framework for the prediction of the mode of transportation using just GPS data is provided by Etemad et al. [25]. The main contribution of the authors' work is their suggestion for the generation of trajectory point features and the extraction of features from trajectory segments, which include bearing rate, bearing change rate, and global and local trajectory features.

The authors then make an effort to categorise the trajectory segments of the moving object according to the various modes of transportation, like walking, riding a train, a bus, or a bike, and driving, with the aid of a number of machine learning techniques, including SVM, decision tree, and XGBoost. Dabiri and Heaslip [26] offer a travel mode inference model based on convolutional neural network (CNN) techniques that are capable of automatically driving high-level features from raw input rather than using handcrafted features with conventional machine learning algorithms.

The authors use raw GPS trajectory data to forecast transport mode labels, such as walk, bike, bus, automobile, and train. The proposed method, which is based on CNN architectures, achieves cutting-edge accuracy when applied to GPS data from the GeoLife dataset [23]. In the work of Zheng et al. [24], the authors suggest a supervised learning-based strategy to infer people's modes of mobility, such as driving, walking, taking the bus, and riding a bicycle.

Starting with a list of features that are more resilient to traffic conditions—such as heading change rate, stop rate, and velocity change rate—the authors define a set of features. The authors then extract features from each segment of the trajectory data and use them to train a supervised learning model using transportation mode inference after segmenting the trajectory data using change point-based detection. Kontopoulos et al. [26] make an effort to categorise vessel trajectories from real-time streams into three activities, which include trawling, longlining, and under way, in another area of work connected to vessel trajectory mining.

The suggested method divides vessel trajectories into temporal segments of 1, 2, 4, 8, 12, and 24 h before generating a set of features that are indicative of the relevant activities, such as the average speed and its standard deviation, the average drift and its standard deviation, and the number of turns. To determine the fishing activities of the vessels, the authors train a random forest classifier and evaluate it against three additional classifiers: gradient boosted tree, linear discriminant analysis, and logistic regression.

11.5.2 Activity Recognition from Wearable Sensors

An active field of research has evolved with the primary objective of extracting information from data collected by pervasive sensors as a result of the remarkable recent development of mobile devices and high-computational, small-sized, and low-cost sensors. Researchers in the field are particularly interested in the task of recognising human activity from wearable sensors because it has a wide range of potential applications, such as tracking athletes' activities and keeping track of patients with diabetes or heart disease as they go about their daily routines. For this reason, it is essential to recognise actions like running, walking, standing up, and raising your hand in order to give input regarding various environmental scenarios. In this section, we examine the prior research on recognising human activity (HAR) from wearable sensors.

The data acquisition mechanism from wearable sensors to the storage in a local device or a remote server is shown in Fig. 11.6. In the beginning, sensors that are

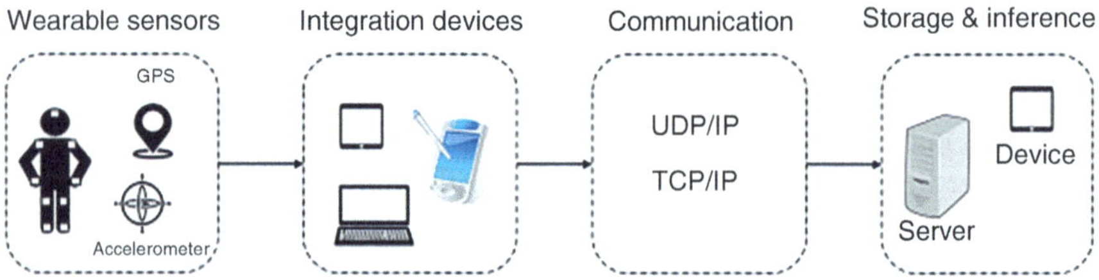

Fig. 11.6 Generic data collecting architecture for human activity recognition

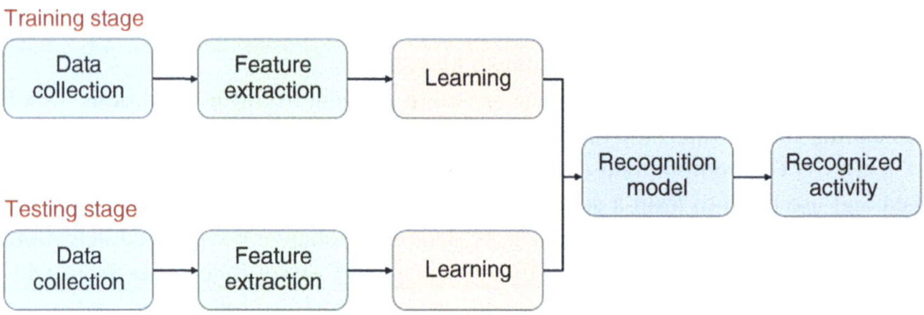

Fig. 11.7 HAR system architecture based on wearable sensors

affixed to a person's body take measurements of pertinent data on many phenomena, including motion [1, 7], location [20], temperature [23], ECG [25], etc.

These wearable sensors can connect to a customised embedded system, a mobile, a PDA, a laptop, or other integration devices. The application server receives the sensor data and sends it for visualisation, analysis, or real-time monitoring. Depending on the desired level of reliability, the communication protocol can be either UDP/IP or TCP/IP. It is important to note that every described component is not used in any HAR system and that the deployment of each component varies on the application scenario.

The authors also provide a generic architecture for all HAR systems. They contend that, like any other machine learning application, activity detection from wearable sensors goes through two stages: training and testing (also known as assessment).

Figure 11.7 depicts the typical processes that these two stages go through. The time series signals are divided into time windows during the training phase, where feature extraction is used and pertinent data is filtered. The extracted feature collection is then utilised to create an HAR model using machine learning techniques.

Similar to this, data was divided up into time windows throughout the testing phase. The same features are extracted from the segmented data and used to assess the previously trained model. Following the same HAR system architecture, Parkka et al. [26] classified commonplace activities like walking, running, and cycling using a variety of data signals obtained from wearable sensors, including an accelerometer, microphone, and air pressure sensor.

Using a 1-s segmentation technique, the authors were able to extract six features from the signal, including the peak frequency of up-down chest acceleration, the median of up-down chest acceleration, and the peak power of up-down chest acceleration. Three distinct classifiers—a custom decision tree, an automatically produced decision, and an artificial neural network (ANN)—were employed to categorise the segments into everyday activities.

Zhang and Sawchuk [138] provide a system based on bag-of-features (BoFs) to construct activity identification models employing motion primitive symbols in another intriguing piece of work in the healthcare evaluation arena. The authors demonstrate the efficacy of their BoF-based methodology for the classification of nine activity classes, including running forward, jumping up, sitting on a chair, standing, and walking forward, left, and right.

11.6 Automatically General Learning

To solve the issue of recognising complicated activities from numerous sensors, such as gestures or actions, algorithms create a dictionary of time series patterns (referred to as shapelets). By redefining the shapelet as a representation of the activity, the authors expand the concept of shapelets to represent complicated activities. Alternately, Jiang et al.'s study [20, 23] proposes to recognise human physical activities from accelerometer and gyroscope signals by learning automatically the best characteristics, as opposed to employing handmade features from sensor signals. The suggested method converts sensor input sequences into pictures and applies deep convolutional neural networks (DCNNs) to identify the best characteristics for distinguishing between actions like walking, standing, and going downstairs. Ordóez et al. [26] developed a framework for activity recognition using convolutional and LSTM recurrent units, staying within deep learning-based ideas.

Their proposal's main selling point is the automatic feature design that does not require specialised knowledge. The authors show how their framework may be used to infer behaviours such as locomotion, postures, and gestures using wearable sensors. In addition, Wang et al. [24] give an overview of current developments in sensor-based deep learning methods for activity recognition.

We discussed two taxonomy systems for activity recognition based on GPS data and wearable sensors in addition to the type of activity. Additionally, we offered a substantial collection of research ideas that use feature extraction and classification to determine the nature of an activity, depending on whether GPS data or signals from wearable sensors are used in the processing. We emphasised that various wearable sensors, including accelerometers, microphones, and air pressure sensors, are utilised in the HAR research ideas.

However, since the information from an accelerometer or sound is private, we do not use them in the context of MCS. Additionally, the earlier abandoned solutions were centred on either spatial or temporal information, but there is a lack of a comprehensive methodological strategy for fusing these many features on real-world enhanced trajectory data. Instead of using a single characteristic, this combination may produce a more reliable inference model. This combination has to be studied, and we will explore it in Chap. 3 of this dissertation.

11.6.1 Generic Machine Learning Algorithms

The previous activity recognition algorithms rely on GPS and wearable sensor data. In reality, the human activity recognition challenge is not limited to these two methods alone, but also applies to additional use cases where multivariate time series are produced from diverse data sources [26]. The study designs that may be used for reasoning process of various human activity extracted from multivariate time series are reviewed under this section. As the two primary generic machine learning techniques, the focus is more on multivariate time series classification (MTSC) and multi-view learning.

11.6.2 Multivariate Time Series Classification

The difficulty in providing labelled data segments with specific activity that results in an MTSC problem based on data gathered from numerous sources is human activity recognition. Numerous time series classification techniques exist, and they can be divided into four groups: deep learning models [27], feature-based methods, ensemble methods [28], and distance-based methods [21]. In order to provide a preliminary assessment in the MTSC problem, the one-nearest neighbour (1-NN) classifier using various distance metrics, e.g. Euclidean distance (ED) or dynamic time wrapping (DTW) [29], is always taken into consideration as the standard to be followed.

When using feature-based approaches, we may distinguish between different types of data and categorise them using a variety of features that have been learned from TS data. The intricacy and weak generality of the building features, which obviously restricts these methods' adaptability, are their drawbacks. These techniques adhere perfectly to the strategy described previously and shown in Fig. 11.8.

Other than manually created features, various techniques extract time series' features for classification using deep neural networks (DNNs). The most recent DNN architectures for time series classification (TSC), including convolutional neural network (CNN), recurrent neural network (RNN), echo state network (ESN), and multi-layer perceptron (MLP), are empirically studied in Fawaz et al.'s survey [48]. The authors tested the methods on multivariate time series datasets in addition to univariate time series, and they provide an overview of the most effective deep learning applications.

Some researches employed both labelled and un-labelled data to obtain knowledge about the human activity, which is known as semi-supervised learning (SSL)

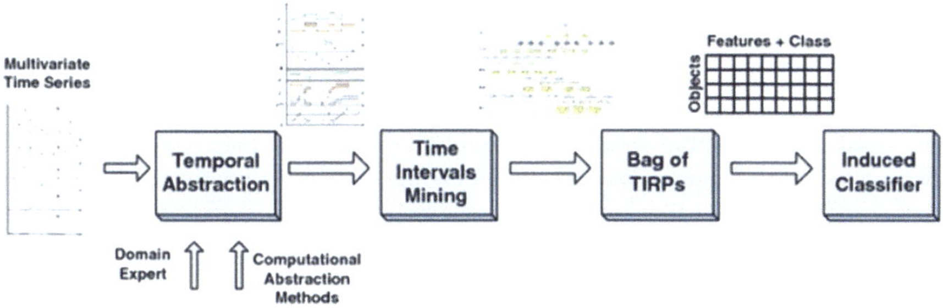

Fig. 11.8 Multivariant time series classification

[25] on MTSC, taking into account the real-life conditions when it is quite challenging to gather voluminous labelled data used for training the algorithm. The groundbreaking work by Alsahfi et al. [25] proposes an SSL method of classifying time series. The authors proved that training on a limited set of labels results in lower accuracy and less human effort than semi-supervised learning.

The one-nearest-neighbour (1-NN) classifier may be considered as a foundation of the semi-supervised model [26], which is founded on the idea of self-learning. The 1-NN classifier C is first trained using the labelled set P (as positively labelled). The pseudo-labels are then gradually applied to the un-labelled sample U defined over its proximity to the provided samples in P. Then, using the augmented labelled set P, it is possible to perform the preceding phase repeatedly while also refining the classifier.

The deep learning-based models on MTSC have lately demonstrated impressive performance with minimal supervision. In order to delve the important facts present in the un-labelled samples, Zhang et al. [27] present an advanced semi-supervised MTSC model called time series attentional prototype network (TapNet). TapNet creates a low-dimensional representation space from the raw MTS data. Additionally, the hybrid long short-term memory (LSTM) and convolutional neural network (CNN) structures used in TapNet enable modelling of the changing interactions and temporal characteristics of MTS, respectively.

11.6.3 Multi-View Learning

Another set of studies suggest using multi-view learning to categorise time series data from several sensors in order to identify user behaviours. In order to recognise human movement utilising wearable technology, Garcia-Ceja et al. offer an approach designed over multi-view learning and layered generalisation. Utilising stacking generalisation, the data from each sensor is integrated into a single "view" [28].

The method uses the view models as input to train a specialised classification model designed over each present view as well as an additional meta-learner. Authors' overarching concept is to mix data from many sensor types to complement one another and so improve recognition accuracy. For feature extraction from various elements of data derived over properties of sequential pattern and visualisation, Wang et al. [23] present system based on deep learning. The authors change the time series into an area graph to mimic how the human brain classifies data based on visualisation. Here, time series are modelled as images using an area graph, and then convolutional neural network (CNN) is applied over that model. In order to extract the features of time series data, they employ widely compliant long short-term memory with an attention mechanism (LSTM-A) neural networks and CNN with attention (CNNA). Sequence features are extracted from the time series using LSTM-A, whereas visual characteristics are extracted by CNN-A. The authors then perform the time series classification challenge using the fusion of features. Even though the method produced encouraging results, more performance was acquired by recent deep learning techniques like InceptionTime [9]. Multi-view discriminative bilinear projections (MDBPs) are suggested by Li et al. [26] for multi-view MTSC.

A multi-view-based dimension (MDBP) reduction technique is used for time series classification. It is the suggested method that tries to extract distinguishing characteristics from multi-view MTS data. MDBP trains unique features by incorporating a unique supervised regularisation and projects multi-view data primarily via view-specific bilinear projections, which retains the temporal structure of MTS to a shared subspace.

In summary, the HAR problem is a multivariate time series classification problem that uses sensor measurements as input and activity labelling as an output. MCS data, on the other hand, is distinguished by its varying feature, indicating that it is derived from various sensor-collected data.

In fact, certain sensors can be down and stop transferring data for hours at a time, which could cause issues with missing data. Because of this, using MTSC is not simple. It goes without saying that a model must be created that integrates data from several types of sensors and can effectively classify data even if one (or more) dimensions are absent.

In addition, the multi-view learning strategy appears to be a potential remedy for the heterogeneity of the employed sensors. According to Ambiciti and Elnahrawy et al. [26, 27], each sensor's data is viewed as a unique "view", and ensemble learning [28] can combine the several viewpoints. The model would be able to infer the final label using the given dimensions if some dimensions are missing.

11.7 Summary

The chapter covers an overview on "environmental spatiotemporal data analytics", which highlights the fundamentals of analytics of spatiotemporal based data and various techniques and types of undergoing identification of change of object from one position to another in a trajectory n-dimensional space. Machine learning algorithm is used in order to extract patterns of change, identified from a considered dataset. The chapter also highlights the time series classification and clustering of spatiotemporal data along with multi-view of the data considered for classification using various deep learning methods.

References

1. R. K. Ganti, F. Ye, and H. Lei. Mobile crowdsensing: current state and future challenges. IEEE communications Magazine, 49(11):32–39, 2011.
2. B. Guo, Z. Yu, X. Zhou, and D. Zhang. From participatory sensing to mobile crowd sensing. In 2014 IEEE International Conference on Pervasive Computing and Communication Workshops (PERCOM WORKSHOPS), pages 593–598. IEEE, 2014.
3. Y. Zheng. Trajectory data mining: An overview. ACM Trans. Intell. Syst. Technol., 6(3), May 2015. ISSN 2157-6904. https://doi.org/10.1145/2743025.
4. Y. Zheng. Trajectory data mining: An overview. ACM Trans. Intell. Syst. Technol., 6(3), May 2015.
5. Y. Zheng, L. Zhang, Z. Ma, X. Xie, and W.-Y. Ma. Recommending friends and locations based on individual location history. ACM Transactions on the Web (TWEB), 5(1):1–44, 2011.
6. B. Languille, V. Gros, N. Bonnaire, C. Pommier, C. Honoré, C. Debert, L. Gauvin, S. Srairi, I. Annesi-Maesano, B. Chaix, et al. A methodology for the characterization of portable sensors for air quality measure with the goal of deployment in citizen science. Science of the Total Environment, 708: 134698, 2020.
7. Y. Zheng, X. Xie, W.-Y. Ma, et al. Geolife: A collaborative social networking service among user, location and trajectory. IEEE Data Eng. Bull., 33(2): 32–39, 2010.
8. L. Pappalardo, F. Simini, G. Barlacchi, and R. Pellungrini. Scikit-mobility: a python library for the analysis, generation and risk assessment of mobility data, 2019.
9. A. Hendawi, J. Shen, S. S. Sabbineni, Y. Song, P. Cao, Z. Zhang, J. Krumm, and M. Ali. Noise patterns in gps trajectories. In 2020 21st IEEE International Conference on Mobile Data Management (MDM), pages 178–185. IEEE, 2020
10. M. Gupta, J. Gao, C. C. Aggarwal, and J. Han. Outlier detection for temporal data: A survey. IEEE Transactions on Knowledge and data Engineering, 26 (9):2250–2267, 2013.
11. A. Blázquez-García, A. Conde, U. Mori, and J. A. Lozano. A review on outlier/anomaly detection in time series data. ACM Computing Surveys (CSUR), 54(3):1–33, 2021
12. H. J. Miller. A measurement theory for time geography. Geographical analysis, 37(1):17–45, 2005.
13. H. J. Miller. Time geography. In Handbook of Behavioral and Cognitive Geography. Edward Elgar Publishing, 2018.
14. C. Parent, S. Spaccapietra, C. Renso, G. Andrienko, N. Andrienko, V. Bogorny, M. L. Damiani, A. Gkoulalas-Divanis, J. Macedo, N. Pelekis, et al. Semantic trajectories modeling and analysis. ACM Computing Surveys (CSUR), 45(4):1–32, 2013.
15. Q. Li, Y. Zheng, X. Xie, Y. Chen, W. Liu, and W.-Y. Ma. Mining user similarity based on location history. In Proceedings of the 16th ACM SIGSPATIAL international conference on Advances in geographic information systems, pages 1–10, 2008.
16. Z. Yan, D. Chakraborty, C. Parent, S. Spaccapietra, and K. Aberer. Semitri: a framework for semantic annotation of heterogeneous trajectories. In Proceedings of the 14th international conference on extending database technology, pages 259–270, 2011.
17. L. Gong, T. Yamamoto, and T. Morikawa. Identification of activity stop locations in gps trajectories by dbscan-te method combined with support vector machines. Transportation research procedia, 32:146–154, 2018.

18. M. Ester, H. P. Kriegel, J. Sander, and X. Xu. A density-based algorithm for discovering clusters in large spatial databases with noise. In Proceedings of the 2nd International Conference on Knowledge Discovery and Data Mining (KDD'96), pages 226–231, 1996.
19. Y. Zheng, L. Liu, L. Wang, and X. Xie. Learning transportation mode from raw gps data for geographic applications on the web. In Proceedings of the 17th international conference on World Wide Web, pages 247–256, 2008.
20. A. Bonavita, R. Guidotti, and M. Nanni. Individual and collective stop-based adaptive trajectory segmentation. GeoInformatica, pages 1–27, 2021.
21. G. Palshikar et al. Simple algorithms for peak detection in time-series. In Proc. 1st Int. Conf. Advanced Data Analysis, Business Analytics and Intelligence, volume 122, 2009
22. M. Abboud, H. E. Hafyani, J. Zuo, K. Zeitouni, and Y. Taher. Microenvironment recognition in the context of environmental crowdsensing. Proceedings of the Workshops of the EDBT/ICDT 2021 Joint Conference, 2841, 2021.
23. A. Ali and J. K. Aggarwal. Segmentation and recognition of continuous human activity. Proceedings IEEE Workshop on Detection and Recognition of Events in Video, pages 28–35, 2001.
24. S. Ali, M. A. Jarwar, and I. Chong. Design methodology of microservices to support predictive analytics for iot applications. Sensors, 18(12):4226, 2018.
25. T. Alsahfi, M. Almotairi, and R. Elmasri. A survey on trajectory data warehouse. Spatial Information Research, 28(1):53–66, 2020.
26. Ambiciti. http://ambiciti.io/, Last accessed May 2022. ISSN 2157-6904. https://doi.org/10.1145/2743025.
27. E. Elnahrawy and B. Nath. Cleaning and querying noisy sensors. In Proceedings of the 2nd ACM international conference on Wireless sensor networks and applications, pages 78–87, 2003
28. S. Aminikhanghahi and D. J. Cook. A survey of methods for time series change point detection. Knowledge and Information Systems, 51:339–367, 2016.
29. T. Hägerstrand. What about people in regional science? Papers of the Regional Science Association, 24:6–21, 1970.

Chapter 12
Future and Research Perspectives of Spatiotemporal Data Management Methods

T. F. Michael Raj, G. Vallathan, Eswaran Perumal, P. Sudhakar, and John A.

Abstract This chapter presents different spatiotemporal data management models, challenges and future direction of spatiotemporal research. The spatiotemporal data model and analysis consists of four main parts such as indexing, query processing, updations and applications. The indexing and data management uses to manage past, present and future (PPF) data. The data management method is generally used to manage PPF methods and also used to manage spatial-based priority management. This chapter presents PPF data, spatial-based priority-based methods and corresponding limitations and suggestions. Finally, different issues and limitations of spatiotemporal methods in general-based, task-based, application-based and implementation-based spatiotemporal data management and future research direction are presented.

Keywords Spatiotemporal data · Data management · Research perspectives Challenges of data management

T. F. M. Raj
Department of Computer Applications, SCLAS, SIMATS Deemed University, Chennai, Tamil Nadu, India

G. Vallathan
Department of Electronics and Communication Engineering, Geethanjali College of Engineering and Technology, Hyderabad, India

E. Perumal
Department of Computer Applications, Alagappa University, Karaikudi, Tamil Nadu, India

P. Sudhakar · J. A. (✉)
School of Computing Science and Engineering, Galgotias University, Greater Noida, India

J. A et al. (eds.), *Spatiotemporal Data Analytics and Modeling*, Big Data Management, https://doi.org/10.1007/978-981-99-9651-3_12

12.1 Introduction

The advanced features of wireless communication and technologies in mobile system include the acquisition of huge amount of data collection on the location-based services (LBS). This is one of the challenging tasks in the data mining-based applications. For example, to increase the accuracy of positioning technologies based on Geographical Positioning System (GPS), continuous updating of moving objects, real-time location-based data prediction, predicting the future trajectory of the car are some of the applications based on wireless communication and advanced features of information technology. From time-to-time, information of the location is one of the key features of these applications.

In the real world, many data objects attributes are represented with location (spatial) and time (temporal). Such spatiotemporal objects are found in many real-time applications such as GIS, environmental data managements and multimedia applications. The traditional relational database is not suitable to manage complex data structure and behavior of data [1]. The spatiotemporal data manage both complex data structure and behavior of data. In the spatiotemporal data models, the spatial data and temporal data have mostly been treated independently. But most of the real-time applications considered the combination of spatial and temporal data. The spatiotemporal database is managed by both space (spatial) and time (temporal) information. A simple way to define the spatiotemporal concepts according to the spatial view is with time interval. The fundamental spatial and temporal aspects should consider for the modeling of spatiotemporal applications [2].

The spatiotemporal examples are: (1) tracking of moving objects based on the given time and location; (2) wireless communication network based on location and timing; (3) indexing the species and road network activities based on location and timing; (4) historical tracking of the plate tectonic activity. This thesis work is based on the updating and indexing of road network activities of moving objects.

Most of the spatiotemporal applications can be categorized into three different types, which are as follows:

1. Applications of moving items or simple objects in continuous motion, such as a navigational system, activity recognition, and etc. In the moving object applications, the objects' positions change simultaneously, but the shape does not change. For instance, a "car" runs along a network of roads, but its shape doesn't change, and its position changes simultaneously.
2. Applications built on discrete object transformations. These applications deal with objects in space whose properties, such as shape and position, may discretely change with respect to time. For instance, in the cadastral information system, "land parcels" or "rivers" can change their locations.
3. Applications can control objects with both continuous motion and form change. For instance, a "storm" is measured as a moving phenomenon in environmental applications that alters its characteristics over time. The organization, querying, updating, and reacting to users, as well as the similarity of moving objects, are some of the major problems with spatiotemporal data.

12.2 Spatiotemporal Data Management Methods

This section presents different spatial and temporal data management methods managing spatiotemporal data. Categorizing multi-dimensional objects of databases has several issues for processing steps which should be considered. The processing steps and issues area are indexing, updation, query processing techniques and, if required, grouping of objects done based on the requirements of users. This chapter presents a brief review of related research in spatiotemporal data techniques along with their merits and demerits. This section gives a brief review of spatiotemporal indexing methods, continuous updations of spatial temporal data, various query processing techniques and clustering or grouping of similar objects.

Effective way of arranging data and faster way of retrieving are used in indexing methods. The indexing techniques are categorized based on the usage of data retrieval such as past, present and future data. The three methods of data processing such as i. spatial information, ii. manage spatial and temporal index methods, and iii. manage all the points in terms of temporal with spatial features [2, 3]. Prediction of data in all points is very difficult because the structure of index is multi-structured and multi-dimensional. Some of the spatial access methods are RT-TREE [4], 3D-R – Tree [5], STR – Tree [6], MTSB-Tree [7], FNR [8].

Indexing methods for managing spatial and temporal data are MR-Tree [9], HR-Tree [10], HR+-Tree [11], MV3R-Tree [12]. The managing temporal and spatial priority multi-dimensional tree methods are TB-Tree [6], SETI [13], SEB-Tree [14], PMR-Quad tree [15], TPR-Tree [16], TPR*-Tree [17], LU Grid [18], RUM-Tree [19], IMORS [20], PTI – Index [21], MODTN Model [22], LGU-Tree [23], FUR – Tree [24], MON [25]. The present research direction is based on indexing methods R^{ppf}-Tree [26], PCFI+-Index [27], BB^x-Index [28], UTR-Tree [29], $STCB^+$-Tree [30], PPFI-Index [31], HTPR*-Tree [32], MSMON [33] and PPFN* [34]. This temporal indexing with indexing the spatial priority is used to manage all the points of future, past and present data. This chapter briefly describes a few spatiotemporal methods and future research directions of temporal with spatial dimensional importance indexing methods. The detailed structure of the important indexing methods is as follows. Detailed structure of the some of the indexing methods are as follows.

Different indexing methods are proposed to manage spatial and temporal data. FUR-Tree [35] means frequent updates R-Tree.

It is used for answering only windows queries. In this tree, insertion, updation and deletion are based on the R-Tree. The FUR-Tree supports a bottom-up approach when doing insertion and deletion operations. Updation in this tree is based on top-down approach, and this approach is inefficient because moving objects are stored in leaf node. This method uses R-Tree, so overlapping problems occur. The PTI indexing method is based on probability threshold indexing method (PTI). PTI is based on one-dimensional R-Tree. In this structure, objects are overlapped because it contributes toward probability threshold queries on uncertainty. The PTI structure representation in Fig. 12.1 and the framework is similar to R-Tree, so this structure has overlapping and dead space problems.

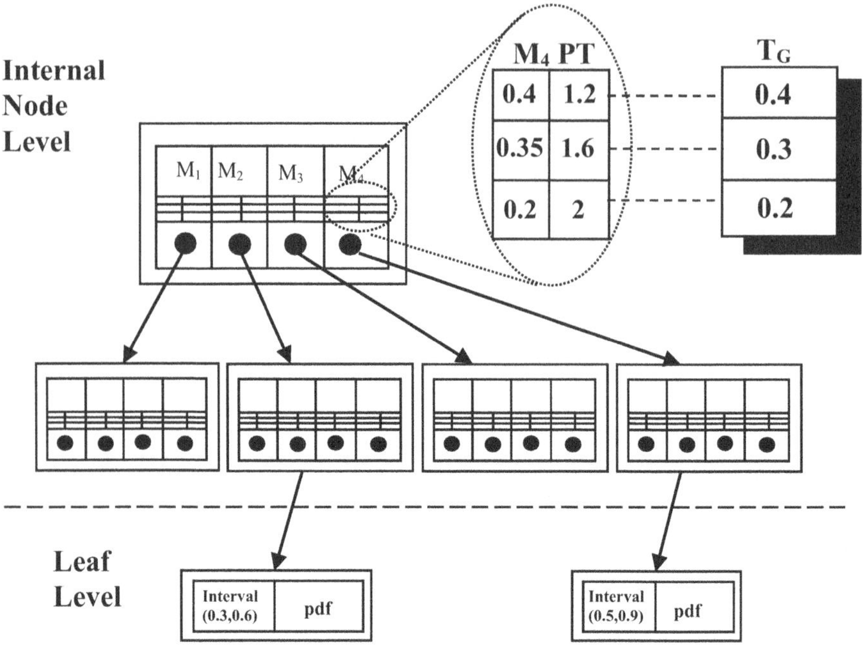

Fig. 12.1 Structure of PTI

Moving Object Dynamic Transport Network model [22] is used for moving objects in pre-defined road network. This model monitors the events of temporary construction of road, traffic jams, deletion of junctions and routes. In the dynamic routes, this model has two steps. One is transportation network and the other is moving graph point in terms of time. The structure of this model is discrete, so it is very complex. Lazy Group Updates (LGU) Method [36] is disk based and it groups similar updates. The two main key roles performed in LGU are disk-based insertion for buffer (I-Buffer) and entire tree structure deletion, which are based on the deletion table (D-Table). In this method, I/O cost is very less from the disk side, but from the user side the I/O processing time is high. The representation of the LGU is given in Fig. 12.2.

12.2.1 Temporal with Spatial Priority Indexing Methods

The temporal with spatial index access methods are used to access past, present and future (PPF) data in multi-dimensional data in terms of future, past and present data prediction. This access method manages and processes the data with spatial and temporal indexing. The important spatiotemporal access methods for future, past, and present are presented in this section. The main indexing trees are PCFI+-Index

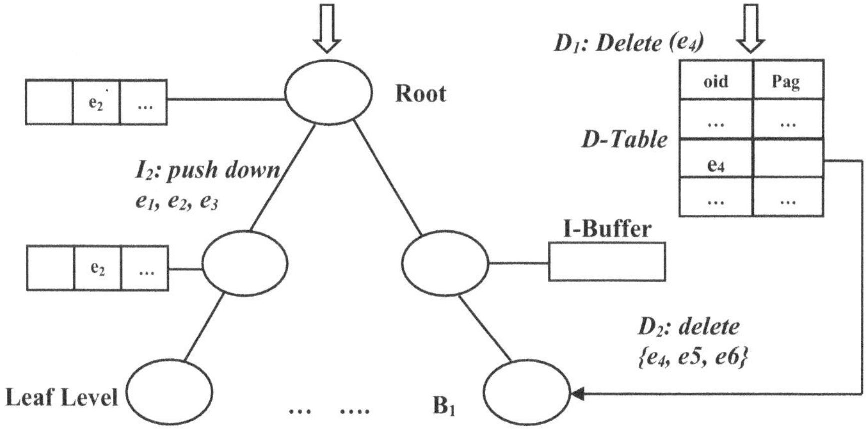

Fig. 12.2 Structure of LGU

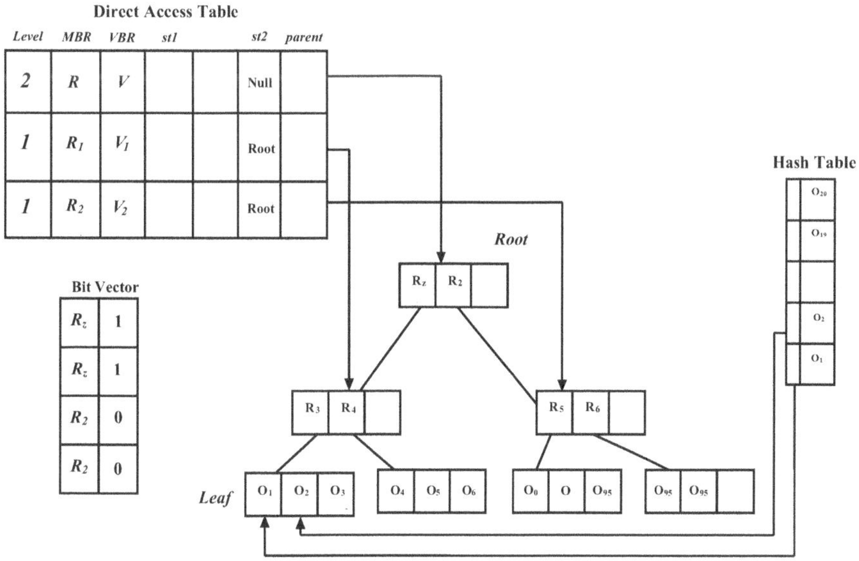

Fig. 12.3 Structure of HTPR*-Index

[27], BB^x-Index [37], UTR-Tree, $STCB^+$-Tree, PPFI-Index, HTPR*-Tree [32], MSMON [33] and PPFN* [38].

The fundamental indexing methods for finding data all the time and all the positions called HTPR*-Tree [32]. The HTPR* means history time parameterized R-Tree which supports predictive queries. The recently updated information is only added to the past data. The recent update is based on the recent time and it is updated from the leaf node. The HTPR* index structure updation is based on R-Tree and it supports the bottom-up approach. The structure of HTPR*-index is shown in Fig. 12.3 and the main advantages of this index is bottom-up approach for pre-defined positions. The new positions and velocity of objects are updated from the top-down approach. The velocity of an object is updated in vector-bounded rectangle (VBR) of root. This index is also updated based on the R-Tree, so overlapping and dead space problems occur.

Another important multi-structure moving objects index in network (MSMON) is used to repossess the future, past and present location of moving objects of road networks in constrained network. The previous structure MON deals only with the recorded data and it cannot perform future and current query. The MSMON tree also indexes the future, past and current information road network and also performs different queries. This index has three features which are (1) moving objects data by dividing time into segments to avoid overlapping and dead space problems. (2) The hash table is used to assist the updates using bottom-up approach. (3) It uses hash table to store all neighbors' information and finally it supports with various queries and predicts all types of information. The structure of MSMON-tree has T3D-tree, NT-Tree and hash tables. The T3D-tree (Time-split 3D R-tree) manages past and historical information. T3D uses two stages: upper stages and lower stages. The upper stage uses R-Tree and lower stage uses hash table. This tree also uses upper stage of R-Tree, so overlapping problems occur in the inner processing part.

The hybrid index method deals with past information and near future information prediction. PPFN*-Tree manages two parts: one is static and the other one is dynamic part; the static part manages the road information. The 2DR*-Tree manages the road records with the help of hash table and TB*-Tree supervises history trajectory besides with past data. The dynamic part is managed by the HTPR*-Tree, and it manages the recent update information.

12.3 Spatiotemporal Updation Methods

Query processing and updation are main issues to be considered in spatiotemporal user query processing. The updation of information in spatiotemporal is based on different scenarios. For example, updation is done with the assistance of various parameters such as speed, location and timing of the moving objects. These parameters are reorganized with the help of various methods. The few methods of updating in spatiotemporal are incremental method, threshold method, PLACE, Min–Max. This section gives brief review about various updation methods which are as follows: The PLACE (pervasive location aware computing environment) was developed by Purdue University [39]. With the help of this continuous spatiotemporal query, they are updated in unbounded areas. The PLACE incremental operators support various spatiotemporal queries. The spatial sliding window queries and shared framework are used for execution of concurrent operations. The incremental

Table 12.1 Comparison of updation methods

Method	Operation techniques	Updation queries
Incremental evaluation	ITA and NTA	Windows queries
PLACE	Incremental	NN-queries
SINA	Incremental, positive and negative updation	K-NN and ANN
Threshold method	Distance-based updation	Point query and window query
CWQ	WOs-filter and WOs-refinement phase	Spatiotemporal queries
CM^2DBQ	Distance	Spatiotemporal queries

evaluation is used to change the last execution data. The main shortcoming of these techniques is that it supports only spatial and NN queries. The different updation methods and corresponding operation techniques and updation queries are summarized in Table 12.1.

12.4 Spatiotemporal Design Requirements

In the implementation of spatiotemporal applications, the following requirements are considered in the processing levels of applications [40]:

1. The requirement for illustrations of objects with points in space and existence in time.
2. Record how a person's position in space changes over time. Two alternative changes of positions are employed depending on the application. (a) Ongoing modifications that cause motion. (b) Discrete changes throughout time, like a change in shape.
3. The requirement for the specification of temporal and geographical attributes in the structuring of layers or files. The visual representation of the spatial and temporal features might be continuous or discrete.
4. Record how spatial characteristics change over time. Changes can happen continuously or abruptly.
5. The need for spatial connections between temporal objects.
6. The requirement for spatial relationships between objects across time to be represented.
7. It is necessary to describe spatiotemporal integrity requirements that are set by the user or the database designer to maintain the database's integrity.

12.5 Issues of Spatiotemporal Data Analytics and Modelling

The different issues and challenges faced while modelling and implementation of spatiotemporal data. This chapter presents four types of issues and challenges for implementation and design. Table 12.2 presents the various supporting issues and challenges related to spatiotemporal data analysis [41].

In general, relationships between the spatiotemporal objects and finding the corresponding relationships, aggregations of the similarities and dissimilarities between the objects are among the main issues in the spatiotemporal data. Similarly, the relationship between the data from one location to another location, segmentation of the regions and relationship between the objects in the spatial environment are the main general challenges in the spatiotemporal data. The task-based issues are one of the main challenges in spatiotemporal data. The different challenges faced in the task-based spatiotemporal environment are as follows. The PPF data positions, analysis and recommendations in location and timing-based predictions. Another important task in spatiotemporal data is the similarity and clustering of data in the static and dynamic environments of spatiotemporal data. The grouping and similar tasks are one of the main challenges because the position of the data and environmental information is temporal, so the analysis and pattern matching will change every minute.

The spatiotemporal applications change the positions and timing in every second. Some applications include location-based analysis and recommendations, traffic analysis and predictions, environmental analysis and changes, tracking of moving objects, location-based social media analysis and smart Internet of Things in location-based applications. These applications are still having huge challenges in the simulations and implementations. Multiagent modeling, STSE, WaveMarker and GeoSim are some tools used to simulate and represent tools [42–45]. Traditionally, data mining and statistics methods are used for the implementation of spatiotemporal data. But, recently artificial intelligence and its subset machine and deep learning have been used for implementation [46].

Table 12.2 Issues and challenges of spatiotemporal data management

General issues	Task-related issues	Application issues	Implementation concepts
Relationships	Prediction	Location-based analysis	Data mining concepts
Aggregation's	Similarity/ clustering	Traffic analysis	Regression
Interdisciplinary	Pattern mining	Environment analysis	Neural network
Data properties	Visualization	Tracking of moving objects	Machine learning
Region discretization	Analysis	Social media analysis	Deep learning
Segmentation of regions	Outlier detection	Smart internet of things	Query processing techniques

12.6 Summary

Spatiotemporal data manage both location and time-based information. The location-based moving object is variant at any time interval in a particular location. The spatial and temporal road networking model represents moving objects in the graph model and assumes that the objects are in linear movement. But the object changes the position with respect to the speed time-to-time. So, the prediction of data is a difficult task. Different methods of prediction and updating of data are proposed in the last few decades. In this chapter, various spatiotemporal data management methods have been presented. The data management methods are considered in spatiotemporal road networking methods: such as spatiotemporal indexing methods for spatial with temporal priority, spatiotemporal updating and spatiotemporal query processing in the road networks.

Further, the localization can be achieved through a wide variety of positioning such as cellular network-based position, GPS and geo-refereed sensor-based positioning or geo-refereed user queries or mobile users access LBSs on their mobile terminals. So, the LBS delivers its services based on the arrangement and predicting data in different situations, continuously updating the data, based on the different position and needs, the updated data is produced according to user requirements and similar data according to the individual user or group of users. So, in future, continuous updation of data with different dynamics features based on the location and timing still required to concentrate on the spatiotemporal data management.

References

1. Wang, Xiaoyu, Xiaofang Zhou, and Sanglu Lu. "Spatiotemporal data modelling and management: a survey."*Proceedings 36th International Conference on Technology of Object-Oriented Languages and Systems. TOOLS-Asia 2000*. IEEE, 2000.
2. Tryfona, Nectaria, Rosanne Price, and Christian S. Jensen. "Chapter 3: Conceptual models for spatio-temporal applications." *Spatio-temporal databases: the CHOROCHRONOS approach* (2003): 79-116.
3. Nguyen-Dinh, Long-Van, Walid G. Aref, and Mohamed Mokbel. "Spatio-temporal access methods: Part 2 (2003-2010)." (2010).
4. Theodoridis, Yannis, Jefferson RO Silva, and Mario A. Nascimento. "On the generation of spatiotemporal datasets." *International Symposium on Spatial Databases*. Berlin, Heidelberg: Springer Berlin Heidelberg, 1999.
5. Theodoridis, Yannis, Michael Vazirgiannis, and Timos Sellis. "Spatio-temporal indexing for large multimedia applications." *Proceedings of the Third IEEE International Conference on Multimedia Computing and Systems*. IEEE, 1996.
6. Pfoser, Dieter, Christian S. Jensen, and Yannis Theodoridis. "Novel approaches to the indexing of moving object trajectories." *VLDB*. Vol. 2000. 2000.
7. Zhou, Panfeng, et al. "Close pair queries in moving object databases." *Proceedings of the 13th annual ACM international workshop on Geographic information systems*. 2005.

8. Frentzos, Elias. "Indexing objects moving on fixed networks." *International symposium on spatial and temporal databases*. Berlin, Heidelberg: Springer Berlin Heidelberg, 2003.
9. Mokbel, Mohamed F., Thanaa M. Ghanem, and Walid G. Aref. "Spatio-temporal access methods." *IEEE Data Eng. Bull.* 26.2 (2003): 40-49.
10. John, A., M. Sugumaran, and R. S. Rajesh. "Performance analysis of the past, present and future indexing methods for spatio-temporal data." *2017 2nd International Conference on Communication and Electronics Systems (ICCES)*. IEEE, 2017.
11. Tao, Yufei, and Dimitris Papadias. "The mv3r-tree: A spatio-temporal access method for timestamp and interval queries." *Proceedings of Very Large Data Bases Conference (VLDB), 11-14 September, Rome*. 2001.
12. Papadias, Dimitris, et al. "An optimal and progressive algorithm for skyline queries." *Proceedings of the 2003 ACM SIGMOD international conference on Management of data*. 2003.
13. Chakka, V. Prasad, Adam Everspaugh, and Jignesh M. Patel. "Indexing large trajectory data sets with SETI." *CIDR*. Vol. 75. 2003.
14. Song, Z., and Roussopoulos N. Seb-Tree. "An Approach to Index Continuously Moving Objects." *The 4th International Conference on Mobile Data Manage-ment, Melbourne, Australia*. 2003.
15. Meng, Xiaofeng, et al. "Moving objects indexing." *Moving Objects Management: Models, Techniques and Applications* (2010): 35-65.
16. Šaltenis, Simonas, et al. "Indexing the positions of continuously moving objects." *Proceedings of the 2000 ACM SIGMOD international conference on Management of data*. 2000.
17. Tao, Yufei, Dimitris Papadias, and Jimeng Sun. "The TPR*-tree: An optimized spatio-temporal access method for predictive queries." *Proceedings 2003 VLDB conference*. Morgan Kaufmann, 2003.
18. Xiong, Xiaopeng, Mohamed F. Mokbel, and Walid G. Aref. "LUGrid: Update-tolerant grid-based indexing for moving objects." *7th International Conference on Mobile Data Management (MDM'06)*. IEEE, 2006.
19. Silva, Yasin N., Xiaopeng Xiong, and Walid G. Aref. "The RUM-tree: supporting frequent updates in R-trees using memos." *The VLDB Journal* 18 (2009): 719-738.
20. Kim, Kyoung-Sook, et al. "Fast indexing and updating method for moving objects on road networks." *Fourth International Conference on Web Information Systems Engineering Workshops, 2003. Proceedings.*. IEEE, 2003.
21. Prabhakar, Sunil, Rahul Shah, and Sarvjeet Singh. "Indexing uncertain data." *Managing and Mining Uncertain Data, Advances in Database Systems* 35 (2009): 299-325.
22. Ding, Zhiming, and Ralf Hartmut Guting. "Managing moving objects on dynamic transportation networks." *Proceedings. 16th International Conference on Scientific and Statistical Database Management, 2004.*. IEEE, 2004.
23. Paz, Petervir. "Communication Network among Small-Scale Miners: Exploration of Narratives on the Mining Information Provision." *The International Journal of Sustainability in Economic, Social and Cultural Context* 19.1 (2022): 53.
24. Lee, Mong Li, et al. "Supporting frequent updates in r-trees: A bottom-up approach." *Proceedings 2003 VLDB Conference*. Morgan Kaufmann, 2003.
25. De Almeida, Victor Teixeira, and Ralf Hartmut Güting. "Indexing the trajectories of moving objects in networks." *GeoInformatica* 9.1 (2005): 33-60.
26. Pelanis, Mindaugas, Simonas Šaltenis, and Christian S. Jensen. "Indexing the past, present, and anticipated future positions of moving objects." *ACM Transactions on Database Systems (TODS)* 31.1 (2006): 255-298.
27. Liu, Zhao-Hong, et al. "Indexing Large Moving Objects from Past to Future with PCFI+-Index." COMAD. 2005.
28. Lin, Dan, et al. "Efficient indexing of the historical, present, and future positions of moving objects." *Proceedings of the 6th international conference on Mobile data management*. 2005.

29. Ding, Zhiming. "UTR-tree: An index structure for the full uncertain trajectories of network-constrained moving objects." *The Ninth International Conference on Mobile Data Management (mdm 2008)*. IEEE, 2008.
30. Tan, Baoquan, Qizhi Liu, and Yunhan Wang. "STORING AND QUERYING HISTORICAL DATA STREAMS BASED ON BD STRUCTURE." *Jisuanji Yingyong yu Ruanjian* 28.2 (2011): 76-79.
31. Fang, Ying, et al. "Indexing the past, present and future positions of moving objects on fixed networks." *2008 International Conference on Computer Science and Software Engineering*. Vol. 4. IEEE, 2008.
32. Fang, Ying, et al. "HTPR*-Tree: An efficient index for moving objects to support predictive query and partial history query." *Web-Age Information Management: WAIM 2011 International Workshops: WGIM 2011, XMLDM 2011, SNA 2011, Wuhan, China, September 14-16, 2011, Revised Selected Papers 12*. Springer Berlin Heidelberg, 2012.
33. Zhu, Zhanyu, Qun Yang, and Dechang Pi. "Past, current and future positions index of moving objects in networks." *Future Control and Automation: Proceedings of the 2nd International Conference on Future Control and Automation (ICFCA 2012)-Volume 1*. Springer Berlin Heidelberg, 2012.
34. Fang, Ying, et al. "Indexing the past, present and future positions of moving objects using PPFI." *2012 8th International Conference on Computing Technology and Information Management (NCM and ICNIT)*. Vol. 1. IEEE, 2012.
35. Mokbel, Mohamed F., Xiaopeing Xiong, and Walid G. Aref. "SINA: Scalable incremental processing of continuous queries in spatio-temporal databases." *Proceedings of the 2004 ACM SIGMOD international conference on Management of data*. 2004.
36. Lin, Bin. *Optimization issues for indexing moving objects*. University of California, Santa Barbara, 2006.
37. Lin, D., et al. "Efficient indexing of the historical, and future positions of moving." (2005).
38. Fang, Ying, et al. "Efficient indexing of the past, present and future positions of moving objects on road network." *Web-Age Information Management: WAIM 2013 International Workshops: HardBD, MDSP, BigEM, TMSN, LQPM, BDMS, Beidaihe, China, June 14-16, 2013. Proceedings 14*. Springer Berlin Heidelberg, 2013.
39. Mokbel, Mohamed F., et al. "PLACE: A query processor for handling real-time spatio-temporal data streams." *Proceedings of the Thirtieth international conference on Very large data bases-Volume 30*. 2004.
40. Pfoser, Dieter, et al. "Dynamic travel time provision for road networks." *Proceedings of the 16th ACM SIGSPATIAL international conference on Advances in geographic information systems*. 2008.
41. Hamdi, Ali, et al. "Spatiotemporal data mining: a survey on challenges and open problems." *Artificial Intelligence Review* (2022): 1-48.
42. Pałka, Piotr, et al. "Using multiagent modeling to forecast the spatiotemporal development of the COVID-19 pandemic in Poland." *Scientific Reports* 12.1 (2022): 11314.
43. Stoma, Szymon, et al. "STSE: Spatio-temporal simulation environment dedicated to biology." *BMC bioinformatics* 12.1 (2011): 1-11.
44. Rohlfs, Wilko, Manuel Rietz, and Benoit Scheid. "WaveMaker: The three-dimensional wave simulation tool for falling liquid films." *SoftwareX* 7 (2018): 211-216.
45. Huang, Jing, et al. "Advancing the application of a model-independent open-source geospatial tool for national-scale spatiotemporal simulations." *Environmental Modelling & Software* 119 (2019): 374-378.
46. Wang, Jing, et al. "Spatiotemporal modeling and prediction in cellular networks: A big data enabled deep learning approach." IEEE INFOCOM 2017-IEEE conference on computer communications. IEEE, 2017.

The manufacturer's authorised representative in the EU is Springer Nature Customer Service Centre GmbH, Europaplatz 3, 69115 Heidelberg, Germany. If you have any concerns regarding our products, please contact ProductSafety@springernature.com

Printed and bound by CPI Group (UK) Ltd, Croydon, CR0 4YY
02/10/2026
02227132-0006